Cambridge Elements ≡

Elements in Flexible and Large-Area Electronics
edited by
Ravinder Dahiya
University of Glasgow
Luigi G. Occhipinti
University of Cambridge

STRETCHABLE SYSTEMS

Materials, Technologies and Applications

Yogeenth Kumaresan
University of Glasgow

Nivasan Yogeswaran
University of Glasgow

Luigi G. Occhipinti
University of Cambridge

Ravinder Dahiya
University of Glasgow

CAMBRIDGE
UNIVERSITY PRESS

CAMBRIDGE
UNIVERSITY PRESS

University Printing House, Cambridge CB2 8BS, United Kingdom

One Liberty Plaza, 20th Floor, New York, NY 10006, USA

477 Williamstown Road, Port Melbourne, VIC 3207, Australia

314–321, 3rd Floor, Plot 3, Splendor Forum, Jasola District Centre, New Delhi – 110025, India

103 Penang Road, #05–06/07, Visioncrest Commercial, Singapore 238467

Cambridge University Press is part of the University of Cambridge.

It furthers the University's mission by disseminating knowledge in the pursuit of education, learning, and research at the highest international levels of excellence.

www.cambridge.org
Information on this title: www.cambridge.org/9781108794824
DOI: 10.1017/9781108882330

© Yogeenth Kumaresan, Nivasan Yogeswaran, Luigi G. Occhipinti and Ravinder Dahiya 2021

First published 2021

A catalogue record for this publication is available from the British Library.

ISBN 978-1-108-79482-4 Paperback
ISSN 2398-4015 (online)
ISSN 2514-3840 (print)

Stretchable Systems

Materials, Technologies and Applications

Elements in Flexible and Large-Area Electronics

DOI: 10.1017/9781108882330
First published online: December 2021

Yogeenth Kumaresan
University of Glasgow

Nivasan Yogeswaran
University of Glasgow

Luigi G. Occhipinti
University of Cambridge

Ravinder Dahiya
University of Glasgow

Author for correspondence: Ravinder Dahiya, ravinder.dahiya@glasgow.ac.uk

Abstract: Stretchable electronics is one of the transformative pillars of future flexible electronics. As a result, the research on new passive and active materials, novel designs and engineering approaches has attracted significant interest. Recent studies have highlighted the importance of new approaches that enable the integration of high-performance materials, including organic and inorganic compounds, carbon-based and layered materials, and composites to serve as conductors, semiconductors or insulators, with the ability to accommodate electronics on stretchable substrates. This Element discusses various strategies that have been developed to obtain stretchable systems, with a focus on various stretchable geometries to achieve strain invariant electrical response. Here, recent advances in material research and their integration techniques towards the development of high-performance stretchable electronics are summarised. In addition, some of the applications, challenges and opportunities associated with the development of stretchable electronics are discussed.

Keywords: electronic devices, Kirigami substrate, meanders, polydimethylsiloxane (PDMS), sensors, stretchable interconnects, stretchable system, structural engineering, wavy and serpentine geometry

ISBNs: 9781108794824 (PB), 9781108882330 (OC)
ISSNs: 2398-4015 (online), 2514-3840 (print)

Contents

1 Introduction

Stretchable systems are required for next-generation functional electronics in applications such as wearable systems, epidermal electronics, soft robotics and electronic skins (e-skins), among others, to allow greater manoeuvrability or to improve user comfort [1–13]. Stretchable systems are required to achieve conformal contact to curvilinear surfaces for real-time monitoring of human health and other environmental updates useful for various applications in healthcare, the military, human–machine interaction, human motion detection and energy harvesting [14–17]. Such applications call for conformable contact with curved surfaces along with a certain degree of stretching and mechanical deformation. As an example, the conformal contact with curvilinear surfaces, enabled by stretchable systems, makes a huge difference in terms of obtaining the reliable physiological data for health monitoring [18]. With high-speed data conversion and data communication, the flexible and stretchable systems are critical for real-time monitoring and the control of various environments with a range of physical, chemical and biological sensors [19–22]. Similarly, flexible and stretchable energy devices are needed in the self-powered systems [23–26].

So far, two major approaches have been commonly utilised to realise stretchable systems: (1) a fully stretchable system, which involves the fabrication of the devices that are intrinsically stretchable and deform in response to mechanical strain; and (2) a semi-stretchable system, which involves the usage of stretchable interconnects to connect the rigid islands of silicon or inorganic semiconductor-based devices (Figure 1.1) [27–32]. These are discussed in detail in Section 2. In fully stretchable systems, the device itself has the capability to handle the strain while stretching. To achieve fully stretchable systems, two major techniques are broadly adopted: (1) direct utilisation of stretchable elastomeric materials; and (2) realisation of bespoke structures using inorganic materials [30, 32–35]. In the former technique, the material itself exhibits the intrinsically stretchable property which is reversible. With multilayer integration of stretchable materials, such as elastomeric conductors, dielectrics and semiconductors, such systems can be deformed to accommodate large strain without any appreciable performance degradation [29]. For example, fully stretchable alternating current electroluminescent (ACEL) devices have been developed using stretchable dielectric nanocomposite consisting of ceramic nanoparticles (NPs) made of $BaTiO_3$ (BT)/poly(vinyl pyrrolidone) (PVP) core–shell NPs dispersed into a polar elastomer, namely a poly (vinylidene fluoride–hexafluoropropylene) (PVDF-HFP) elastomer [30]. As shown in Figure 1.1a, the ACEL device using stretchable active materials demonstrates high mechanical deformability with the ability to tightly integrate

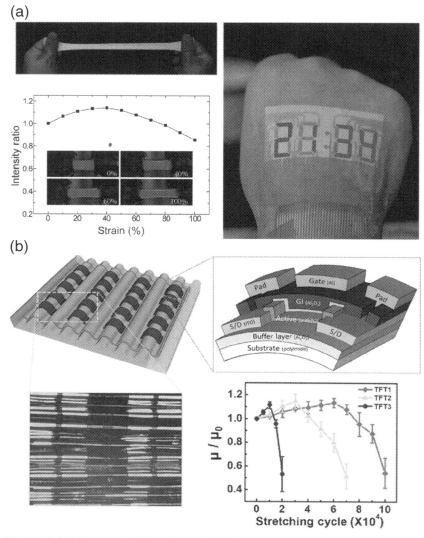

Figure 1.1 Fully stretchable and semi-stretchable systems. (a) Fully stretchable ACEL devices using elastomeric dielectric nanocomposites consisting of BaTiO$_3$/PVP core–shell NPs embedded in a PVDF-HFP elastomer. Adapted with permission from [30]. Copyright (2019) American Chemical Society. (b) Intrinsically stretchable high-performance inorganic device with wavy In-Ga-Zn-O thin-film transistors (IGZO TFTs) on ultrathin polyimide (PI) film/pre-strained elastomer with an outstanding mechanical stretchability. Reprinted with permission from [32]. Copyright (2019) American Chemical Society. (c) A conceptual illustration of a semi-stretchable system with rigid islands that contain an array of electronic components which were bridged or

(c)

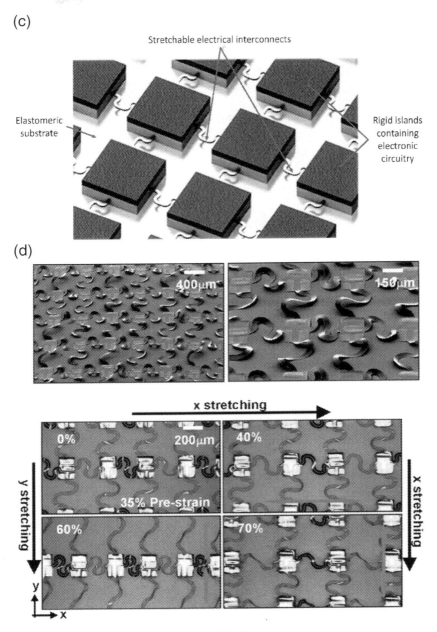

(d)

Figure 1.1 Cont.
interconnected by (e.g., horseshoe-shaped) free-standing metal interconnects. Reproduced from [27]. The overall stretchability of the circuit is mainly determined by the free-standing interconnects.(d) The scanning electron microscope (SEM) image of a bridge–island configuration-based

Caption for Figure 1.1 (cont.)

semi-stretchable system consisting of an array of single crystalline complementary metal oxide semiconductor (CMOS) inverters islands with noncoplanar bridges that have serpentine layouts and the optical images from stretch tests in the x and y directions. Adapted from [28]. Copyright (2008) National Academy of Sciences.

with the human body for epidermal stopwatch application. However, moderate performance of elastomeric materials, including low conductivity of the elastomer and a poor carrier mobility of the stretchable semiconductors, blocks their usage in high-performance applications. In the latter technique, the micro/nanostructures based on inorganic materials in 'wavy' layouts have been adopted for high-performance devices that could offer promising results, as shown in Figure 1.1b, where the amplitude and wavelength is shown to change in response to the applied strains [32, 34].

The semi-stretchable systems introduce stretchability through an island–bridge configuration that contains high-performance inorganic electronic devices as rigid islands and stretchable interconnects as bridges [36–41]. Unlike a fully stretchable system, in which intrinsically stretchable devices accommodate the stretching deformation, the applied stress on the semi-stretchable system is concentrated on the stretchable interconnect itself (Figures 1.1c and 1.1d) [27, 28]. Therefore, it is essential to develop reliable stretchable conductors that meet the stretching requirements. In general, good electrical conductivity and mechanical stretchability are the key parameters needed for stretchable interconnects to bridge the gap between traditional rigid and the next-generation fully stretchable technologies [42]. The mechanical stretchability of interconnects is achieved either by engineering the shape of highly conductive non-stretchable materials or by using intrinsically stretchable conductive elastomers such as block copolymer elastomer conductors (BECs) or ionic conductive elastomers [43–47]. The intrinsically stretchable conductive elastomers can have high reversible deformations owing to their elastic nature, but their high electrical resistance, in comparison with the metal-based interconnects, is a limiting factor for many practical applications [48, 49]. On the other hand, the engineered metallic structures-based stretchable interconnects have very low resistances and low power loss [50, 51]. Over the past few decades, many engineered stretchable structures such as serpentine, fractal, mesh and microcrack structures, among others, have been reported with conductive thin metal films placed on intrinsically stretchable elastomeric substrates, namely polydimethylsiloxane (PDMS), natural rubber, Ecoflex and styrene butadiene rubber (SBR), and so on [52–54]. The choice among these stretchable interconnect schemes ultimately depends on the trade-off between the degree of their stretchability and the resistance. To exploit the best of both of these parameters, alternative strategies such as stacking engineered metallic structures with conductive polymers have also been explored [47]. These alternative solutions provide better conductivity in low stretching regimes and medium conductivity in high stretching regimes, where conductive polymers bridge the cracks in the metallic interconnects.

In this Element, we discuss the recent progress in stretchable systems covering various materials, stretching mechanisms with various geometries and finally the applications. This Element is organised as follows. Section 2 presents the detailed study related to the technologies for stretchable systems on elastomeric substrates. The subsections describe various stretchable geometries such as a wavy structure, a bridge–island structure, and serpentine, fractal, mesh, and microcrack structures, and so on. Section 3 presents different materials utilised for the realisation of stretchable systems. These include organic materials, inorganic materials, elastomeric substrate, stretchable conductors, liquid metal and composites. Section 4 presents fabrication methods and their limitations in terms of integration on elastomeric substrate, and the fabrication techniques for stretchable systems. The printing technique, three-dimensional (3D) writing technique and direct pattern transfer on elastomer substrate are also described. Section 5 presents some exemplar applications of stretchable systems in emerging areas such as wearable systems for health monitoring, e-skins and energy storage. Finally, the key take-home messages are summarised in Section 6 along with future directions for stretchable systems.

2 Stretchable Geometries

Stretchable electrodes or conductors are the basic building blocks of stretchable systems that, as briefly discussed, are realised by engineering the shape or geometry of the various organic and inorganic materials on elastomeric substrate [14, 55]. High-performance inorganic and organic materials (discussed in Section 3) are either rigid or flexible, but they are not stretchable (stretchability is limited to less than 2%). To enable the stretchability, structural design of the material or device is inevitable [56]. As shown in Figure 2.1, the stretchable systems based on inorganic and/or organic materials are broadly classified into two categories as: (1) a fully stretchable system – consisting of intrinsically stretchable devices; and (2) a semi-stretchable system – consisting of a rigid device connected with stretchable interconnects (bridge–island configuration).

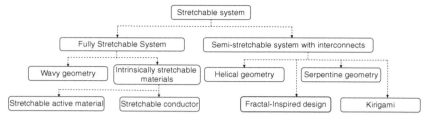

Figure 2.1 Broad classification of stretchable systems and their structural geometries

2.1 Fully Stretchable System

A fully stretchable system requires both the stretchable device and stretchable interconnects to achieve strain invariant device response (semi-stretchable system – only interconnects are stretchable), with the ability to conformally attach on to soft and non-linear surfaces for wearable application. Here, the stretchable devices are presented under the fully stretchable system (Section 2.1) and the stretchable interconnects are presented under the semi-stretchable system (Section 2.2). The stretchability in electronic devices is attained through two techniques: (1) mechanically guided structural design for high-performance inorganic and organic devices; and (2) intrinsically stretchable elastomers or nanowires (NWs) and elastomer composites [56].

2.1.1 Mechanically Guided Structural Design: Wavy Geometry

The fully stretchable systems using organic, inorganic and silicon-based devices are often achieved by wavy (buckled) geometry or through micropatterning [57–62]. To realise a wavy geometry, the rigid thin-films/devices are either directly deposited or transferred to the pre-stretched elastomeric substrate and then simultaneously relaxed to achieve controlled sinusoidal geometry configuration [63–68]. This method is called the mechanical buckling method (MBM), in which the wavelength and amplitude of the sinusoidal wave, responsible for stretchability, are governed by energy methods [14, 69–71].

Accordingly, when the applied pre-strain (ε_{pre}) is less than 5%, a small strain energy model is adopted to investigate the buckling process. The wavelength (λ_o) and amplitude (A) of sinusoidal wave configurations are represented as

$$\lambda_o = 2\pi h \left(\frac{\overline{E_f}}{3\overline{E_s}}\right)^{1/3} \tag{2.1}$$

$$A = h\sqrt{(\frac{\varepsilon_{pre}}{\varepsilon_{min}} - 1)} \qquad (2.2)$$

Where, h is thickness of stiff thin-film, E_f is the Young's modulus of rigid thin-film, E_s is the Young's modulus of substrate and ε_{min} is the minimum strain required to provoke buckling [70–74]. Based on Eqs (2.1) and (2.2), it is evident that the wavelength of the sinusoidal wave, for less than 5% pre-strain, is dependent on the film thickness and not on the pre-strain. Further, the thickness of the film is directly proportional to the wavelength and amplitude, but the pre-strain affects the amplitude of the sinusoidal wave, not the wavelength.

When a large pre-strain is applied (5–30%) on the elastomeric substrate, the wavelength of the sinusoidal configuration increases linearly with applied pre-strain, as shown in Figures 2.2a and 2.2b [70]. In this case, the elastomeric substrate non-linear deformation and non-linear constitutive model is considered [73]. Accordingly, the wavelength is redefined as

$$\lambda = \frac{\lambda_o}{(1 + \varepsilon_{pre})(1 + \xi)^{1/3}} \qquad (2.3)$$

where, λ_o is the wavelength equation for small pre-strain and ξ is $5\varepsilon_{pre}(1 + \varepsilon_{pre})/32$. In agreement with Eq. (2.3), when silicon nanoribbons are bonded to the pre-strained PDMS substrate, the wavelength of the MBM-based silicon nanoribbon sinusoidal waves decrease linearly with the applied pre-strain (Figure 2.2a).

When the MBM-based silicon nanoribbons are subjected to the external stretching strain, the wavelength of the geometry increases and the amplitude decreases [73]. Whereas, under compressive strain, the wavelength decreases while the amplitude increases, as shown in Figure 2.2b [70]. The MBM-based silicone nanoribbons demonstrate the capability to handle up to 29% stretching strain when the PDMS substrate is 29% pre-strained [70]. Based on this strategy, stretchable devices, such as TFTs, energy devices, light-emitting diodes (LEDs) and solar cells, and so on, have been developed [59–61, 63, 64]. For example, zinc oxide (ZnO)-based TFT was transferred to the pre-stretched PDMS to achieve 5% stretchability [63]. To achieve good mechanical durability in solar cells, wrinkled organic solar cells have been fabricated to handle stretching strain and cross-buckled silicon solar cells have been developed to handle both stretching and compressing strain (Figure 2.2c) [59, 65]. Likewise, organic TFTs with a pentacene active layer have been directly fabricated on buckled PDMS substrate and the device demonstrated stable performance up to 12% strain (Figure 2.2d) [59].

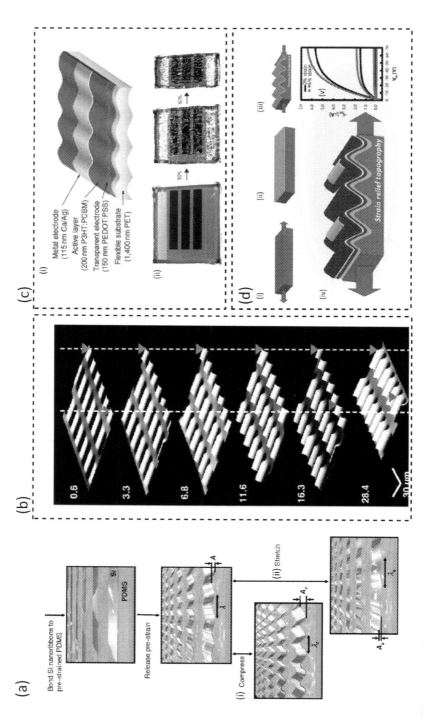

Figure 2.2 Fully stretchable system using sinusoidal wavy geometry. (a) A schematic of the fabrication process for buckled Si ribbon on PDMS substrate achieved by bonding the Si nanoribbon on pre-strained PDMS substrate (top frame) and subsequently releasing the applied

Caption for Figure 2.2 (cont.)

strain (second frame); (i) the compress and (ii) stretch frames illustrate the structural deformation, the amplitude and wavelength of buckled sinusoidal Si NW geometry under compressive strain and stretching strain, respectively. (b) Atomic force micrographs of buckled Si ribbons achieved, by MBM, on different pre-strained PDMS substrate (0–28.4%). The vertical line and the downward triangles illustrate a systematic decrease in wavelength of buckled Si ribbons with respect to the pre-strained PDMS substrate. Adapted with permission from [70]. Copyright (2007) National Academy of Sciences. (c) Stretchable organic photovoltaic (OPV) solar cells; (i) schematic representation of buckled OPV solar cells on elastomer substrate; (ii) optical image of a wrinkled/buckled OPV device achieved by transferring ultra-thin OPV to 100 μm-thick pre-strained elastomer – the optical images at flat (left), 30% (middle) and 50% (right) quasi-linear compression state. Reproduced from [64]. (d) A schematic fabrication process flow of stretchable organic thin-film transistors with wavy configuration. Fabrication steps include (i) uniaxial pre-strain PDMS substrate; (ii) thermally evaporate SiO_2 layer; (iii) release the pre-strain to achieve wavy configuration; (iv) conformably coat the device layer including dielectric layer, the active layer and the source/drain electrodes; and (v) transistor output characteristics under 0 and 6% strain. Reprinted from [59]. Copyright (2013), with permission from Elsevier.

2.1.2 Intrinsically Stretchable Devices

The stretchability in the intrinsically stretchable devices is achieved through material properties that can be controlled through modified polymeric side chains or by placing electronic fillers such as organic/inorganic NWs/flakes inside the elastomeric body [75, 76]. In the case of stretchable organic devices, the side chain engineering has significantly enhanced the device performance and the stretchability through controlled interchain interactions. For example, the stretchable transistors made of isoindigo–bithiophene backbone with branched alkyl side chain-based active material have demonstrated stable electrical performance with the mobility in the range of $\sim 10^{-2} cm^2 V^{-1} s^{-1}$ up to 100% strain [77]. By replacing carbon with silicon in the side chains of tunable linear carbon spacer groups in isoindigo-based polymers, the merits of the carbosilane side chain and the isoindigo-based polymer backbone structure have been utilised to achieve stretchability along with enhanced mobility [78]. Figure 2.3a shows the fabrication of stretchable field-effect transistors (FETs) using stretchable semiconducting polymer and their device performance under various stretching strain. The fabricated FET endured 100% strain with the mobility $>1 cm^2 V^{-1} s^{-1}$. Similarly, stretchability and self-healing properties can be enabled by introducing dynamic non-covalent cross-linking sites on the polymer backbone [79]. The stretchable semiconducting polymer, consisting of 3,6-di(thiophen-2-yl)-2,5-dihydropyrrolo[3,4-c]pyrrole-1,4-dione (DPP) as repeating units with 2,6-pyridine dicarboxamide (PDCA) as a dynamic bonding, has been used to introduce hydrogen bonding in the polymer backbone to significantly increase the tensile modulus of the material [75]. Figure 2.3b shows the stretching mechanism in conjugated polymer with dynamic bonding, in which the DPP repeating units are connected through the PDCA via hydrogen bond strength under stretching. Accordingly, the backbone modification and side chain engineering enhanced the stretchability of the conjugated polymer. The detailed review of such intrinsically stretchable semiconducting polymers for stretchable electronics applications is reported elsewhere [80]. In the case of NWs/flakes in the elastomer matrix, the buckling and interlocking of NWs/ flakes helps to maintain the electrical property under extreme stretching conditions. For example, a rubbery transistor was developed using a stretchable electrode, with semiconductor and dielectric realised by embedding the NWs in elastomeric matrix [81]. The stretchable electrodes were prepared by embedding silver (Ag) NWs inside PDMS and then depositing gold (Au) NPs at the top surface. Likewise, a stretchable semiconductor was realised by embedding poly(3-hexylthiophene-2,5-diyl) (P3HT) nanofibres (NFs) in PDMS with a carbon nanotube (CNT) percolation path for high mobility. Based on these

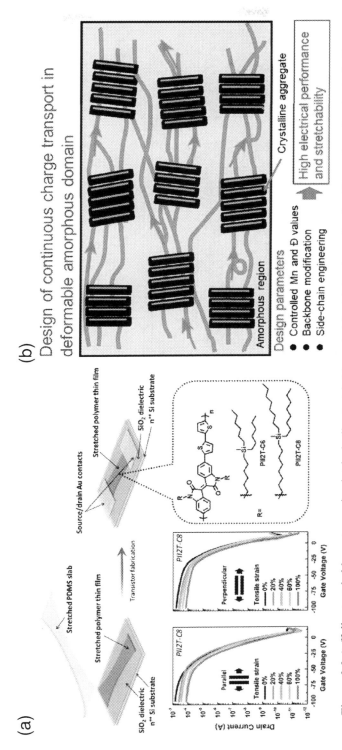

Figure 2.3 A ffully stretchable system using intrinsically stretchable materials. (a) A schematic of an intrinsically stretchable FET device using isoindigo-based conjugated polymers, PII2T-C8, with carbosilane side chains, its chemical structure and the transfer characteristics of a transistor under different stretching conditions (0–100% strain along channel length and width); the FET demonstrated stable performance with a slight decrease in mobility and on-current value under stretching. Reprinted from [78]. Copyright (2016) American Chemical Society. (b) Schematic representing the stretching mechanism of a conjugated polymer semiconductor through

Caption for Figure 2.3 (cont.)

amorphous chain and aggregate via dynamic bonding. Reprinted from [80]. Copyright (2020) with permission from Elsevier. (c) Photomicrograph of an array of rubbery transistor fabricated using m-CNT–doped P3HT-NFs/PDMS semiconductor, Au NPs-Ag NWs/PDMS electrode and ion gel dielectric layer with using m-CNT-doped P3HT-NFs/PDMS semiconductor scheme depicting the charge transport mechanism in the active layer. Reprinted from [81]. Copyright 2019 distributed by CC BY-NC. (d) Schematic representing the stretchable transient conductor from Ag flakes and gelatin hydrogel composite material and the photomicrograph of composite material under stretching and releasing. Reprinted from [82]. Copyright (2020) American Chemical Society. (e) Photomicrograph of graphene oxide (GO)-Ag NW/PUA electrode and its application in stretchable polymer light emitting electrochemical cell (PLEC) with a demo displaying the working of PELC at 130% stretching strain. Reprinted from [83]. Copyright (2014) American Chemical Society. (f) Schematic illustrating an all carbon-based intrinsically stretchable transistor with use of CNT as electrodes and channel layer with styrene-ethylene-butadiene-styrene (SEBS) as substrate and dielectric layer. The plot demonstrates the variation in mobility with respect to the change in diameter of CNT channel. Reprinted from [84]. Copyright (2017) American Chemical Society.

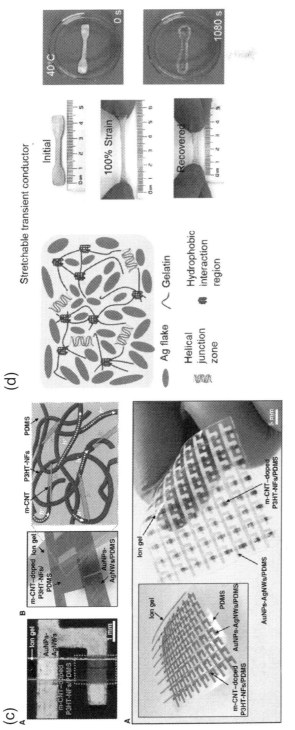

Figure 2.3 Cont.

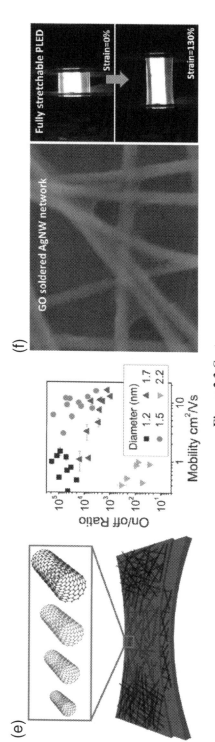

(e)

(f)

GO soldered AgNW network

Fully stretchable PLED

Strain=0%

Strain=130%

On/off Ratio

Mobility cm²/Vs

Diameter (nm)
■ 1.2 ▲ 1.7
● 1.5 ▼ 2.2

Figure 2.3 Cont.

stretchable materials and the ion gel dielectric layer, a stretchable transistor array was developed, as shown in Figure 2.3c. The transistor demonstrated 50% stretchability with the decrease in mobility from 7.46 ± 1.37 to 3.57 ± 1 cm^{2}V^{-1}s^{-1} [81]. In the same way, stretchable transient conductors using Ag flakes and gelatine hydrogel have been developed, as shown in Figure 2.3d [82]. The design shows numerous interaction sites between Ag flakes and hydrogels through strong interfacial bonds, which promotes good mechanical behaviour. The fabricated conductors demonstrated 100% stretchability with good durability above 1000 cycles at 20% strain. Such conductors are used as a sensor for finger joint motion detection and the device demonstrated a reliable response when the sensor was subjected to cyclic 90°C bending through the finger motion.

The intrinsically stretchable behaviour has also been observed when the NWs are adhered on the surface of the elastomeric substrate rather than embedding inside the elastomer matrix. For example, transparent conducting electrodes are obtained by fabricating an Ag NW network on the surface of polyurethane (PU) acrylate stretchable substrate which are soldered using a GO sheet, as shown in Figure 2.3e [83]. The soldering of Ag NW network through GO significantly reduced the inter-wire contact resistances and enhanced the stretchability. Such transparent conducting electrodes have been used for the fabrication of stretchable polymer LED. Likewise, all carbon transistors using single-walled CNT (SWNT) as an electrode and semiconductor layer and SEBS-based nonpolar elastomer as a dielectric layer have also been developed [84]. The variation in the device performance, demonstrated by tuning the diameter of the CNT, is shown in Figure 2.3f. The stretchable transistor with a CNT active layer revealed the mobility of around 25 ± 5 cm^{2}V^{-1}s^{-1} and endured 60% stretching strain. The details regarding the intrinsically stretchable materials are discussed later in Section 3. Along with the stretchable devices, stretchable interconnects are utilised to enhance the stretchability of the device array or integrated devices to realise a stretchable system.

2.2 Semi-Stretchable System-Stretchable Interconnects

Another strategy to realise stretchable electronics is a semi-stretchable system, which utilises bridge and island configuration. In this configuration, the rigid electronic devices are bridged through stretchable interconnects to achieve reversible stretchability. The structural geometry of interconnects plays a vital role in determining the stretchability of the semi-stretchable system. Various structural geometries are briefly summarised in Figure 2.4 [50, 57]. The geometries of stretchable interconnects are broadly classified as helical, serpentine,

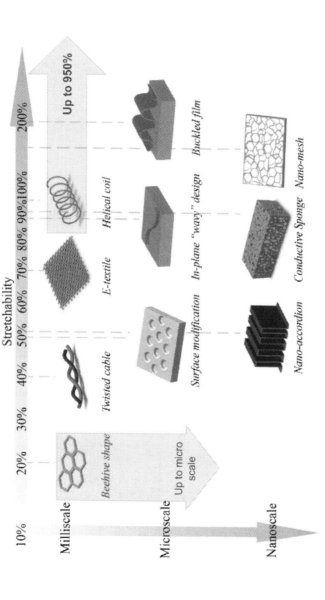

Figure 2.4 Various stretchable conductors/interconnects, designed by employing geometric engineering on non-stretchable materials, sorted according to their elastic deformation with respect to their design dimensions in the range of millimetre to nanometre scale. Reproduced with permission from [50].

fractal-inspired design and Kirigami design. In general, the stretchable geometries are fully bonded to the substrate, for the realisation of in-plane deformation, or partially bonded to the substrate, to have out-of-plane deformation [56].

2.2.1 Stretchable Helical Interconnects

Thin non-stretchable metal wires in the shape of a helical coil or a spring are broadly utilised as interconnects to bear the maximum strain, as shown in Figure 2.5a [85]. The stretching mechanism behind this helical coil is straightforward and depends on the radius of curvature of the coil, width of the material and the pitch between each coil. While stretching, the radius of curvature of the coil decreases and simultaneously the pitch of the coil increases to accommodate the strain. From finite element analysis (FEA), the maximum strain on a unit cell of the spiral geometry is given by:

$$\varepsilon_{12} = \frac{W}{2R - W} \tag{2.4}$$

where, W is the width of the conducting wire and R is the radius of the inner circle [86]. From Eq. (2.4), the W in the denominator has the tendency to decrease the total value of the denominator $(2R - W)$. For larger R, the maximum strain is directly proportional to the width and inversely proportional to the radius. Accordingly, the stretchable geometry should maximise the radius to width ratio to maximise the stretchability. Based on this principle, 3D helix geometry, two-dimensional (2D) spiral shapes, 2D serpentine geometries, fractal inspired designs, twisted cable design and mesh geometries have been developed [55, 87–92]. For example, 3D helical structured interconnects embedded in a polymer substrate (eHelix-Copper (Cu)) have been utilised to realise stretchable helix interconnects capable of handling 150% strain without significant degradation [55]. The 3D helical structure is obtained by fabricating the serpentine-shaped interconnects, between the active component, on a biaxially pre-strained elastomer

and sequentially releasing the partial strain (Figure 2.5a). During this process, the 2D serpentine-shaped interconnects are converted to 3D helical geometry. The 3D helical structure can be completely encapsulated with soft elastomeric material for practical applications, as shown in Figure 2.5a [85]. Likewise, commercialised fibre wound around a screw could lead to controlled helical geometry and by coating the fibre with polymer/Ag nanocomposites, it is possible to obtain the stretchable conductive interconnects. Such Ag PU-

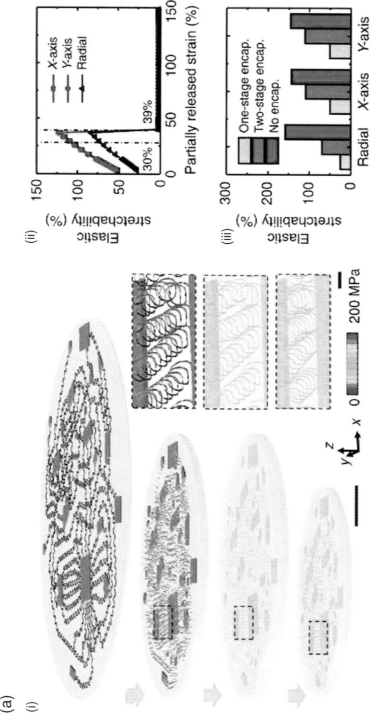

Figure 2.5 Helical- and serpentine-shaped stretchable interconnects. (a) Schematic diagram of fabrication process flow of soft encapsulated 3D helical interconnects with FEA results: first, (i) 2D serpentine interconnects are fabricated on a pre-strained elastomer, the strain is partially released to achieve 3D helical geometry, encapsulated with soft elastomer layer and finally the strain is released; (ii) elastic stretchability of 3D helical structure is encapsulated by a soft elastomer at a different state of partially released strain; and (iii) elastic

Caption for Figure 2.5 (cont.)

stretchability of the 3D helix at a different encapsulation stage. Reproduced with permission from [85]. (b) The schematic illustration of a normal and expandable state of spiral interconnects fabricated using thin copper wire wrapped/wound around nylon fibre. Reproduced with permission from [96]. Copyright (2011), with permission from Elsevier. (c) Digital image of a stretchable sweat pH sensor and radio-frequency identification (RFID) antenna fabricated using serpentine-shaped stretchable interconnects. Adapted from [97]. Copyright (2018), with permission from Elsevier. (d) Theoretical concept of stretchable serpentine routing by considering θ, W/R and L/R of the serpentine design. Adapted from [98].

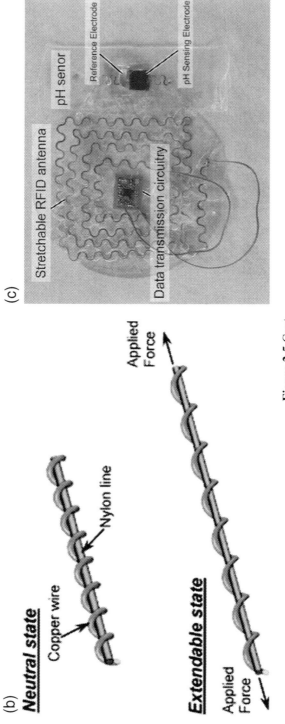

Figure 2.5 Cont.

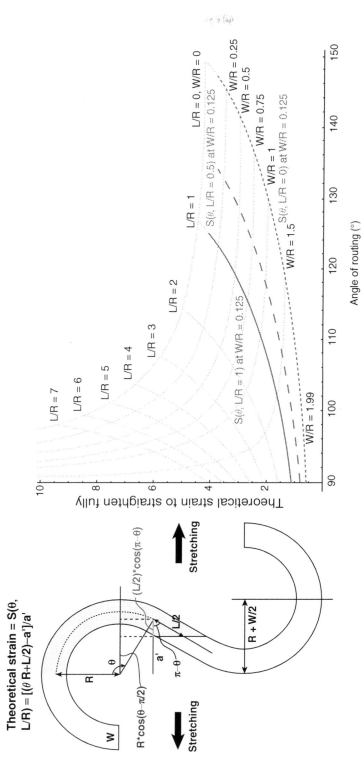

Figure 2.5 Cont.

based helical conductive fibres exhibit remarkable electrical performance under stretching [87]. Similar results have been reported for a wide range of wearable applications [90, 93–5].

Another example of helical-shaped stretchable interconnects is the anthropomorphic robotic skin with large-area tactile sensor array, obtained using expandable spiral geometry achieved by winding the copper wire around an elastic nylon line, as shown in Figure 2.5b [96, 99]. In most of the spiral geometries, the core material is required to guide the pathway of the helical wire's movement, and the conducting wire is wrapped around with precise pitch control (Figure 2.5b) [96]. The selected conducting material can either be a flexible polymer nanocomposite fibre or rigid conducting copper wire [87, 96]. Based on material selection (either inorganic or organic), winding pitch and winding dimension, such interconnects can be reversibly stretched up to 950% strain with minimum stress on the conducting material to retain its electrical performance [87].

2.2.2 Stretchable Serpentine Interconnects

Another engineering strategy for realising the stretchable interconnects involves the utilisation of serpentine-shaped configuration [100–105]. The scalability that is the main disadvantage of helical-shaped geometry can also be addressed by using serpentine-shaped interconnects, which can be easily scaled down to microscale by microfabrication techniques [105, 106]. Serpentine geometry can be considered as the extended version of the spiral geometry and therefore the maximum strain on the serpentine geometry is expressed as

$$\varepsilon_{12} = \frac{6W}{R - W} \tag{2.5}$$

where R is the radius of the half circle and W is the width of the conducting geometry. Like the helical geometry, the maximum strain in the serpentine geometry is directly proportional to the width and inversely proportional to the radius of the semicircle (Eq. (2.5)) [86]. While stretching, the curved surface of the serpentine geometry undergoes bending distortion to redistribute the normal stress. Figure 2.5c shows the stretchable system with the use of serpentine interconnects [97]. Several mathematical models have been developed to study the working mechanism and to optimise the stress distribution in serpentine geometry [103, 104, 107]. As an example, the model based on the space-fill principle utilises the long interconnects in patterned geometry along certain regions with the estimation of higher stretchability [104]. The boundary conditions in this case are defined by clamping both the ends of interconnects. However, this model does not consider the interaction between the interconnects and the substrate, which makes it less reliable. For an accurate estimation, another model considering the plastic deformation in interconnects has been proposed. In this model, the serpentine-shaped interconnects are modelled as an Euler–Bernoulli beam with both the ends clamped [107]. Similarly, a simple model using serpentine geometry is proposed, as shown in Figure 2.5d, by considering the radius of arc (R), arc angle (θ), straight section of pattern (L) and the width of the pattern (W) [98]. The strain-influencing factor to straighten fully with respect to the angle of serpentine routing by considering different L/R and W/R ratio is given in the Figure 2.5d. However, material failure and interaction failure between substrate and conducting were not taken into consideration [98].

Analytical model and numerical simulations, such as finite element method (FEM) and FEM simulation software (COMSOL), are normally performed to accurately analyse the stress distribution mechanism in serpentine-shaped geometry and to design the stretchable interconnects [108]. The comparison of different serpentine geometries named as 'SerpXX' (XX = θ), triangle geometry, strain line and their elongation ratios is given in Table 2.1. For an accurate estimation, material-dependent qualities such as Lames parameters have been considered in the model and the boundary conditions are defined with perfect bonding between the interconnects and PDMS substrate to match the experimental sequence. The material parameters in the simulation model, composed of 100 nm-thick Au metal interconnects and 0.16 mm-thick elastic PDMS modelled as hyper-elastic polymer (Neo-Hookean model), correlate with the experimental result. The solid mechanics module and AC/DC module of COMSOL have been utilised here to define the mechanical movement and to monitor the resistance change in interconnects, respectively. Based on the simulation, the concentrated von Mises stress and location of maximum stress in different interconnects were shown in Figures 2.6a–2.6f. The concentrated von Mises stress is located at the crest and trough of the serpentine, and the apex of triangular interconnects. Likewise, in the strain line interconnects, the maximum stress is located at the ends. The resistance value and the variation in resistance for various stretchable geometries under 70% strain are given in the Figures 2.6g and 2.6h, respectively. The resistance variation is strongly related to the stress that determines the stretchability of the interconnect, so, stretchability of the different serpentine- and triangle-based geometries derived from their resistance value are as follows: Serp260 > Serp45 > Triangle > Serp180 > Serp60 > Straight line [108]. The simulated resistance and von Mises stress for Serp260 agree with the geometric elongation but this is not true for Serp180.

Table 2.1 Comparison of geometric elongation ratio of straight line, serpentine design with different angle (θ) and triangular geometry. Adapted from [108].

Name	Schematic	Geometrical elongation ratio
Straight line	Electrodes for clamps 0.5 mm 10 mm	0
Serp 45	45°	33.4%
Serp 60	60°	3.7%
Serp 180	180°	55.6%
Serp 260	260°	180%
Triangle	90°	34.5%

Based on the geometric elongation, Serp180 should demonstrate higher stretchability than the triangle and Serp45. This study indicates that along with the geometric deformation, material properties also play an important role in terms of deciding the stretchability of the interconnects.

2.2.3 Other Stretchable Geometries

The stretchability of helical or serpentine geometries is limited to a uniaxial direction (the stretchability along the X-direction is more than 50 times greater compared with the Y-direction). Such shortcomings can be addressed by using the serpentine pattern in a triangular geometry (fractal design) to connect the sensors placed on a rigid island [109]. An example in Figure 2.7a shows the supercapacitor connected with the periodic interconnects designed using fractal-inspired geometry to demonstrate the omnidirectional stretchability with minimum strain acting on the interconnect metal layer. The fractal-inspired-design-based stretchable interconnects provide a high coverage area with enhanced omnidirectional stretchability by filling the 2D lines with curved patterns [104]. Such designs can be obtained by utilising the normal serpentine or rectangular geometry as a first-order structure and higher fractal orders by

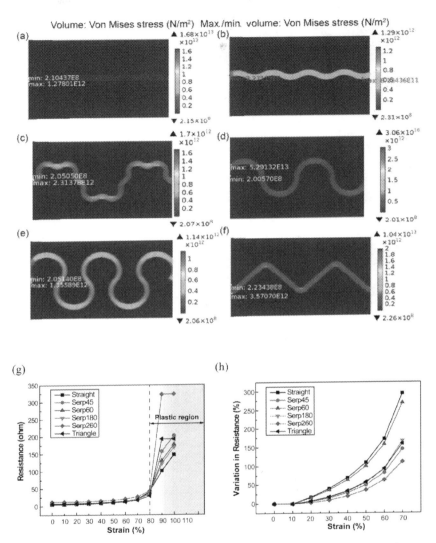

Figure 2.6 (a–f) Comparison of stress distribution, such as concentrated von Mises stress and the location of maximum von Mises stress, within various geometries such as straight line, triangle and different serpentine-shaped interconnects under 70% stretching. Adapted from [108]. (g) The plot between resistance value with respect to the stretching strain for various geometries with highlighted plastic region above 80% strain indicating non-linear behaviour of various geometries, and (h) the change in resistance of various interconnects up to 70% stretching. Adapted from [97]. Copyright (2018), with permission from Elsevier.

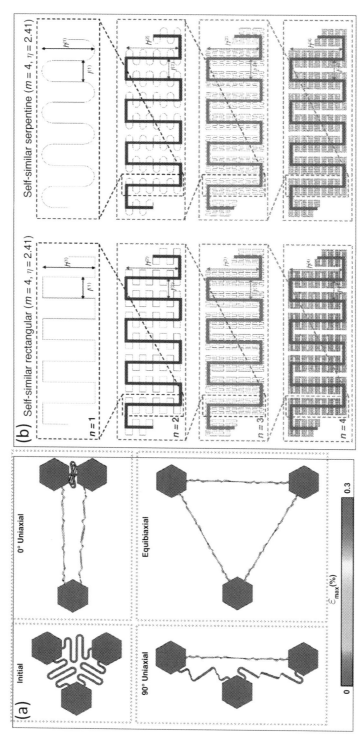

Figure 2.7 Fractal-inspired design. (a) A schematic illustration of biaxial elongation of rigid electronic device (supercapacitor unit cell) array in hexagon geometry connected by self-similar fractal-inspired serpentine bridge design (initial), demonstrating the change in interconnects geometry while stretching along uniaxial (0° and 90°) and biaxial directions. Adapted with permission from [109]. (b) The design of fourth-

Caption for Figure 2.7 (cont.)

order self-similar rectangular and serpentine interconnects (bottom), along with the magnified design geometry, clearly demonstrates different orders such as third, second and first (bottom to top), respectively, and (c) the stretchability as a function of self-similar order. Reproduced from [104]. Copyright (2013), with permission from Elsevier. (d) A digital image of stretchable solar cells fabricated by using the Kirigami pattern-inspired stretchable interconnects. Adapted with permission from [36]. Copyright (2019) American Chemical Society. (e) The stretching mechanism of Kirigami pattern investigated by creating a systematic line cut pattern, with the cut length greater than the spacing, and their mechanical response namely Force, F, with respect to extension, Δ, are shown. Three regions are observed: initial linear regime with in-plane deformation; second regime with out-of-plane bending; and final regime with deformation localised near to the edges of the cut. Reproduced with permission from [110].

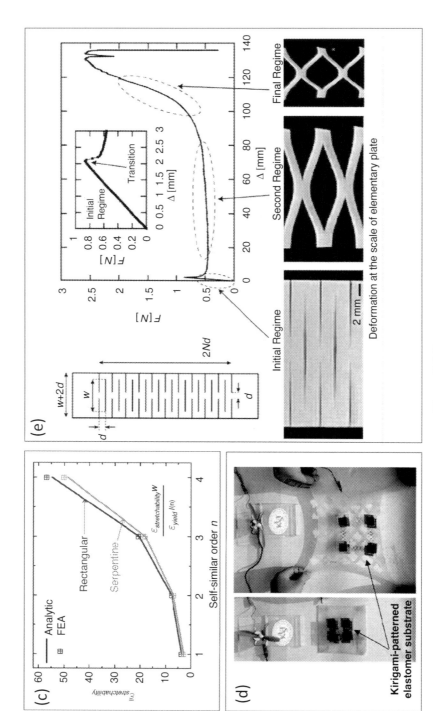

Figure 2.7 Cont.

filling the 2D line in the first-order structure with similar geometry at different levels to have second-, third- and fourth-order geometries, as shown in Figure 2.7b.

The analytical and FEM simulation models can be developed to study the stretching mechanism of such space-filling fractal designs with respect to their fractal orders. As an example, it has been reported that the elasticity of such stretchable interconnects can increase more than twice with the order of a similar structure increasing by one (Figure 2.7c), thus offering an extreme level of biaxial stretchability, up to 90% [104]. Similarly, many other geometries such as square, zigzag, honeycomb, nano-mesh, microcracks, substrate-induced design and Kirigami structures are also used as interconnects or stretchable conductors for the stretchable system [40, 111–20]. Kirigami designs (inspired by Japanese paper cut art, where 'Kiri' is cutting and 'gamai' is paper) are also utilised to realise stretchable systems [40, 112–22]. Figure 2.7d demonstrates an omnidirectional stretchable solar cell using Kirigami-based stretchable interconnects [36]. The Kirigami-inspired 3D pattern is encapsulated here in an Ag NW network arrangement for perovskite solar cells with an aerial coverage of 97% and a system stretchability of 400% and a 25,000% interconnect stretchability. To understand the stretching mechanism of Kirigami-based patterns, the relationship between the mechanical response with respect to the pattern elongation can be explored. As shown in Figure 2.7e, the mechanical force can be plotted as a function of pattern extension to obtain non-linear mechanical behaviour with three different deformation regimes – that is, from an initial linear deformation regime to secondary elastic deformation and finally the pattern collapse regime [110]. After reaching the peak at the initial linear regime, there is a transition from 2D to 3D deformation in the Kirigami pattern [110, 122]. The softness of the Kirigami pattern emerges from the transition. The strain is equally distributed on the Kirigami pattern under stretching, as a result, despite multiple defects, the tolerance of the Kirigami-based interconnect is improved and offers strain-independent electrical conductance in the interconnects.

Substrate topography also plays an important role in enhancing the stretchability. In this regard, the conductive materials are directly coated on the engineered substrate, such as a micropatterned, mogul-patterned elastomer or taro-leaf-templated PDMS substrates, to enhance the biaxial stretchability [123– 6]. The strain distribution on the metal layer coated on top of the mogul-patterned elastomeric substrate consists of a bump/valley structure. It can be analysed by experiment and computer simulation. Based on both analyses, it is found that under applied strain, the stress is more concentrated in the valley region than in the bump. Accordingly, a metal layer is connected continuously along the bump region to develop stable electrical performance under 50% biaxial stretching [124]. In addition to this bump/valley-based substrate, the honeycomb-designed substrate or sponge-shaped substrate can also be used for stretchable interconnects [127, 128]. The conductivity in sponge-shaped materials can be achieved either by sputtering or through drop-casting the conductive materials, namely Ag or CNT/graphene networks [129–31]. For example, the stretchable metallic Ag nanoporous sponge embedded in polytetrafluoroethylene was developed using sputtering to achieve 50% stretchability [128]. The 3D porosity in sponge-like materials could effectively redistribute the strain acting on the conducting layers by rotational movement to maintain electrical property under stretching. However, the sponge-shaped materials have several limitations such as challenging integration with the stretchable system, and there is a trade-off between the stretchability and the scalability. In general, fine pore size in sponge is preferable for device scalability but this affects the mechanical durability. To overcome such limitation, nano-mesh geometries are developed [132]. As an example, the metal nano-mesh patterns created on rigid silicon, using the grain-boundary lithography technique, and transferred to the elastomeric substrate, can lead to 160% stretchability.

3 Materials for Stretchable Interconnects

Materials are key components for the development of stretchable electronics; therefore, the appropriate material must be chosen to render the final system with the degree of stretchability required for the target application. The choice of material is also important in terms of alignment with the fabrication technologies as several materials are incompatible with the available manufacturing processes. This is due to challenges such as rigidity of materials or the harsh fabrication conditions (e.g., high temperature, pressure and chemicals) that are often incompatible with soft materials used in the development of flexible and stretchable electronics. This has fuelled substantial research into the development of novel materials as well as device layouts and integration to enhance the stretchability. In particular, smart structural engineering has been proven to be a successful approach in integrating the conventional monocrystalline inorganic materials, thereby allowing the realisation of high-performance electronic devices. This section provides an overview of various materials that have been employed in the development of stretchable electronics devices. The key terms discussed include substrates, conductors, semiconductors and an overview of structural engineering employed for the development of stretchable components.

3.1 Substrate Materials

Substrates are critical components in the realisation of electronic devices; therefore, it is necessary that a substrate with inherent stretchability is chosen as the building block for the development of stretchable electronics. In this regard, elastomers have been widely explored as substrates for the development of stretchable electronics. The inherent flexibility of the elastomers has been explored in various conformable electronics applications such as e-skins [133] and stretchable sensors [134], and so on. Besides mechanical properties, the biocompatibility and optical transparency of elastomeric substrates have resulted in their use in applications such as biomedical implants [97] and optoelectronic devices (LEDs, transparent electrode and photodetectors, etc. [135–7]).

Silicone-based elastomers such as PDMS, Ecoflex and silicone rubber are widely used substrate materials in the development of stretchable electronics [49, 138, 139]. Along with an intrinsic stretchable property, the low cost, facile processing techniques and ease with which its surface and elastic properties can be modified have made it a popular material for the stretchable electronics application. In general, three types of crosslinking, namely chemical or physical or both chemical and physical, take place between polymer chains in the elastomer network. In the case of PDMS and Ecoflex elastomers, chemical crosslinking takes place where the polymer chains are connected through strong covalent bonds. The mechanical properties of PDMS can be tuned by controlling the crosslinking and modulus in different ways, which includes: (1) varying the composition ratio of the prepolymer and the curing agent [140]; (2) adding chemicals [141]; and (3) varying the curing temperature [142]. The surface properties of PDMS such as its hydrophobicity can be temporarily altered by O_2 plasma or ultraviolet (UV)/ozone [143, 144] and chemical treatment. Such temporary modification attained by the aforementioned treatments is harnessed to improve its adhesion with material such as thin metal films, and so on. Further, the self-healing property in PDMS is achieved through coordination bonds by crosslinking the PDMS polymer with different Zn(II)-diiminopyridine coordination complexes, as shown in Figure 3.1a. Thus, mechanically weaker and more dynamic bonds are created due to the presence of side methyl groups, which significantly enhance the stretchability and introduce a self-healing property [145]. The PDMS crosslinked by Zn(II)-diiminopyridine coordination complexes named as a PDMS-MeNNN-Zn polymer demonstrate high stretchability, with its breaking point at 456% strain. Based on a rheological recovery test at 25°C (Figure 3.1a), the PDMS-MeNNN-Zn polymer breaks partially at 500% strain, demonstrates 83.7% recovery within 10 s and 90% recovery after

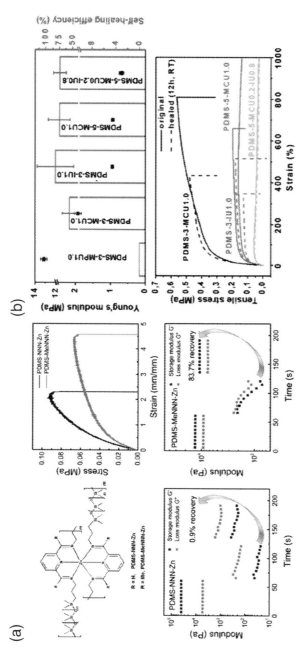

Figure 3.1 Intrinsically stretchable substrate materials and their functionalisation to enhance the elastic modulus. (a) The chemical structure of PDMS polymers cross-linked by two different Zn(II)–diiminopyridine coordination complexes and their mechanical and self-healing properties from a rheological recovery test. Reprinted from [145]. Copyright (2018) American Chemical Society. (b) Tuning the elasticity and self-healing property of PDMS by different IU, MCU and MPU mixing ratios. Reprinted from [146]. Copyright (2020) American Chemical Society. (c) Schematic diagram of Electrospun SEBS/PANi micro-nano-fibre membranes. Reprinted from [150]. Copyright (2020) with permission from Elsevier. (d) Optical photograph of mesh heater made from ligand exchange (LE) Ag NWs/SBS composite material encapsulated by an SBS layer. Adapted from [151]. Copyright (2015) American Chemical Society. (e) Summary of the mechanical properties of a PU elastomer mixed with various organic or hybrid fillers. Adapted with permission from [154]. Copyright (2017) American Chemical Society. (f) A self-healable elastomer. A schematic illustration at the top and demonstration of self-healing of the device using LED light. Reproduced from [158] with permission from the Royal Society of Chemistry.

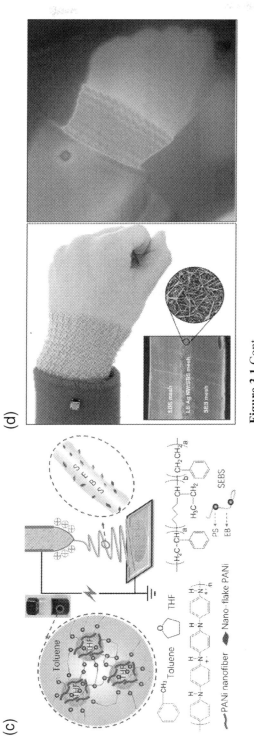

Figure 3.1 Cont.

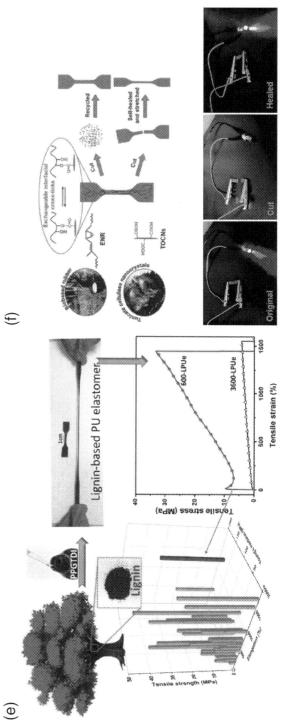

(e)

(f)

Figure 3.1 Cont.

30 min at room temperature. Similarly, isophorone diisocyanate (IU), 4,4'-methylenebis(cyclohexyl isocyanate) (MCU) and 4,4'-methylenebis(phenyl isocyanate) (MPU) have been used with a PDMS precursor at different mixing ratios to achieve supramolecular interactions capable of reversible breakage and reformation for a self-healable application [146]. Accordingly, minimum recovery of 80% in stress is achieved by optimising the ratio of MPU and MCU against PDMS, as shown in Figure 3.1b. In certain applications requiring biocompatibility such as non-invasive bio-monitoring applications, new elastomers such as Silibione silicone rubber have recently been explored [147–9].

Styrene-based elastomers such as poly(styrene-butadiene-styrene) (SBS) and poly-SEBS are examples of physically crosslinked elastomers in which the polymer chains are connected through weak interactions, such as hydrogen bonding, hydrophobic interactions and chain entanglements, rather than a strong chemical connection [76]. Styrene-butadiene-styrene and SEBS are thermoplastic elastomers consisting of a polystyrene domain with soft butadiene or an ethylene–butylene elastomeric matrix. The weak polymer chain interaction due to a physical crosslink in the thermoplastic elastomer results in a good processability needed to achieve any desirable shape by controlling the external pressure or heat. In addition, the thermal stability can be enhanced with the filler materials. Based on these properties, the SEBS/PANi NF membranes have been achieved using an electrospinning technique, with PANi NF electrostatically crosslinked on SEBS, as shown in Figure 3.1c. The fabricated SEBS/PANi micro-nano-fibre membrane demonstrated high stretchability of more than 300% tensile strain and enhanced thermal stability [150]. Furthermore, the physically crosslinked elastomers have an advantage of handling a high stretching strain and simultaneously allowing maximum loading of filler to retain the property of filler. Accordingly, homogenously conductive elastomers have been obtained by 40% Ag NW loading into an SBS network, which is not possible with a chemically crosslinked elastomer as the high loading affects the polymerisation process [151]. As shown in Figure 3.1d, the wrist heater is demonstrated using the combination of a densely loaded Ag NW–SBS composite in an SBS substrate for continuous thermotherapy. Another styrene-based elastomer is SBR, which consists of two monomers randomly distributed in a polymer chain. It is not a thermoplastic. The SBR is often used as a soft stretchable substrate, with mixing of an electronic filler such as CNT/GO inside to enhance the thermal conductivity needed for temperature sensing [152].

Polyurethane is another material that has been explored as the substrate for the stretchable electronics, which contains both a physically and chemically crosslinked network offering higher stretchability [153]. A comparison between stretchable substrate materials and their mechanical properties (elongation and Young's modulus) is given in Table 3.1. PU has been explored in the development of the stretchable printed circuit board [153], but the low Young's modulus and stiffness limited its application [154]. To enhance the stiffness, the PU is mixed with various organic polymers or hybrid fillers (CNT and graphene). For example, the stiff and biodegradable lignin is blended with poly(propylene glycol)tolylene 2,4-diisocyanate terminated (PPGTDI) to form lignin-based PU elastomers (LPUes) with enhanced mechanical properties (a Young's modulus of 176.4 MPa, tensile strength of 33 MPa and elongation of 1,394%) [154]. The property of PU elastomers is easily tuned by varying the lignin molecular weight (Figure 3.1e) [154]. Hydrogels also contain both physically and chemically crosslinked polymer networks. The stretchability and Young's modulus of hydrogels can be adjusted by modifying the polymer content in the hydrogels, which has significant advantage for biomedical applications [155]. Further, the shape-changing property in a hydrogel elastomer is easily achieved by mixing it with the suitable filler. This is useful in various applications, such as soft actuators, bio sensors and controlled drug release. For example, TiO_2 cross-

Table 3.1 Comparison of various stretchable substrate materials and their mechanical parameters

Materials	Elongation	Young's modulus	Ref.
PDMS	~100%	0.8–3 MPa	[159]
Ecoflex	600–980%	1–300 kPa	[160]
Dragon skin	507%	500 kPa	[161]
Blupren PPI90	350%	~ 22 kPa	[139]
Silibione silicone rubber	220%	3 kPa	[148]
PU rubber	375%	6 MPa	[162]
LPUes	1,394%	176.4 MPa	[154]
SBR	380%	2.9 MPa	[152]
SBS	850%	0.5 MPa	[163]
SEBS	452%	23.28 MPa	[164]
TiO2 cross-linked nanocomposite hydrogels	2,492%	4 kPa	[156]
Self-healable substrates			
TEMPO-mediated oxidation of CNs	400–800%	2–10 MPa	[158]
PDMS-MPU0.4-IU0.6	1,600%	0.7 MPa	[157]
PDMS-MeNNN-Zn	456%	0.067 MPa	[145]
PDMS-5-MCU0.8-IU0.2	1,034%	0.71 MPa	[146]

linked nanocomposite hydrogel demonstrates directional shape switching useful for soft actuator application [156]. Similarly, self-healable materials such as
2,2,6,6-tetramethyl-piperidinyl-1-oxyl radical (TEMPO)-mediated oxidation of
cellulose nanocrystals (CNs) or PDMS-MPU0.4-isophorone bisurea unit (IU)
0.6 have been explored to mimic the nature of human skin for various wearable
and prosthetic applications [157, 158]. In self-healable materials, dynamic
covalent chemistry is adopted in which the covalently cross-linked network
has the potential to reassemble under external stimuli such as thermal reaction.
As shown in Figure 3.1f, a conducting electrode on a self-healable elastomer
connected to an LED light demonstrates a healable property after being dis-
sected into two pieces [158].

3.2 Stretchable Conductors

A conductor is a vital part in the development of any electronic devices and
circuits as electrical contacts, electrodes or interconnects. Besides the high
electrical conductivity, conductors used in the development of stretchable
electronics should preferably have printing compatibility, exhibit a negligible
variation in its electrical conductivity under strain, be compatible with the
various substrates and have a low-processing temperature. Commonly used
electrical conductors for the stretchable electronic applications include thin
metal films, metallic nanomaterials (NWs and NPs), carbon-based nanomaterials (CNT, graphene or carbon black), liquid metal and polymer conductors. An
overview of various materials used in the development of the stretchable
conductors is discussed here.

3.2.1 Thin Metal Films

The preliminary work on the development of stretchable conductors exploited
the formation of buckling of a thin metallic film on elastomeric surfaces such as
PDMS due to compressive stress (shown in Figure 3.2a) [165]. Such compressive stresses have been explored to develop stretchable Au electrodes on PDMS
substrate [166]. For example, thin Au film strips (100 nm-thick) deposited on
PDMS via a polyimide shadow mask have been shown to withstand up to 22%
strain. Further, by forming 'island-like cracks' in an Au film on an elastomer
(Figure 3.2b), while maintaining a percolation pathway, it is possible to produce
a conductive film capable of maintaining a stable resistance over 250,000 cycles
[167]. The maximum strain the film can withstand can be improved by careful
geometric structuring of film into various shapes such as serpentine [97, 168].
Besides electrodes, geometrical engineering has also been explored for stretchable RFID antennae [97]. For example, the stretchable high conductance

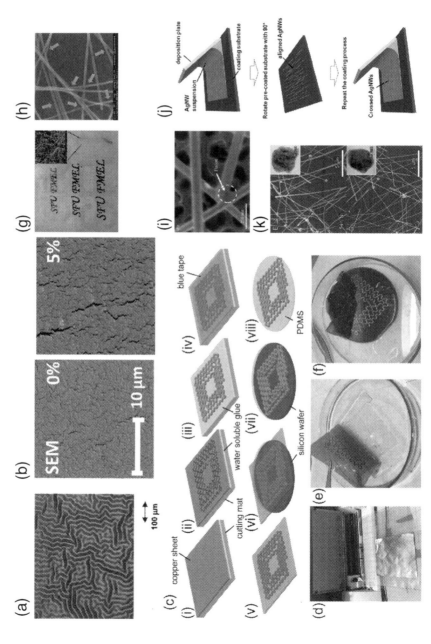

Figure 3.2 Stretchable conductors and applications. (a) Optical micrography of a waveform of thin Au film deposited on PDMS. Reprinted from [166] with the permission of AIP publication. (b) An SEM image of a corresponding island-like crack formation on Au film deposited

Caption for Figure 3.2 (cont.)

on PDMS at 0 and 5% strain. Reprinted from [167] with the permission of AIP publication. (c) A schematic illustration of the process steps involved in the fabrication of stretchable antennae: (i) placing flexible copper sheet onto a cutting mat with a water-soluble glue as sacrificial layer; (ii) blade cutting the antenna pattern; (iii) removing the rest of material; (iv) attaching the blue tape over the antenna pattern; (v) dissolving the glue in water and transferring the antenna pattern to blue tape; (vi) adhering to partially cured PDMS substrate (separately prepared through spin-coating on silicon wafer); (vii) curing PDMS and dissolving blue tape in acetone; and (viii) detaching the PDMS substrate from the silicon wafer. Photos of critical steps c(i), c(v) and c(vii) are also included in (d), (e) and (f), respectively. Reprinted from [97] with permission from Elsevier. (g) Transparent Ag NWs spray deposited on a dopamine-modified PDMS substrate. An SEM image shows the homogenous distribution of Ag NWs on the PDMS surface. Reprinted with permission from [178]. Copyright (2012) American Chemical Society. (h) GO soldered Ag NW junction, which is indicated by an arrow mark. Reprinted with permission from [83]. Copyright (2014) American Chemical Society. (i) An SEM image of annealed NWs fusing together with neighbouring NWs [183, 184]. (j) A schematic illustration of the meniscus dragging process and the SEM image of parallel and cross-Ag NW thin-film formed by this process. Reprinted with permission from [185]. Copyright (2016) American Chemical Society. (k) Copper NWs before (top image) and after coating with nickel (Ni) (bottom image). The inset shows a TEM cross-section image of Cu NWs. Reprinted with permission from [186]. Copyright (2012) American Chemical Society. (l) A piezoresistive Ag NW–PDMS composite for strain sensor application; (i) different electrical interaction between adjacent NWs; and (ii) change in resistance under different bending angle (a), cyclic response (b), motion detection (c) and control of avatar figure using wireless smart glove (d). Reprinted with permission from [187]. Copyright (2014) American Chemical Society. (m) Carbon nanotube-incorporated porous elastomer composite-based conductors for pressure sensor application and their sensor characteristics demonstrating the change in current with respect to various pressure and different loading/unloading cycles. Reprinted with permission from [188]. Copyright (2019) American Chemical Society. (n) A bimode sensor using porous carbonyl iron particles/multiwalled CNT–polydimethylsiloxane composites (PCMCs). Reprinted with permission from [189]. Copyright (2018) American Chemical Society.

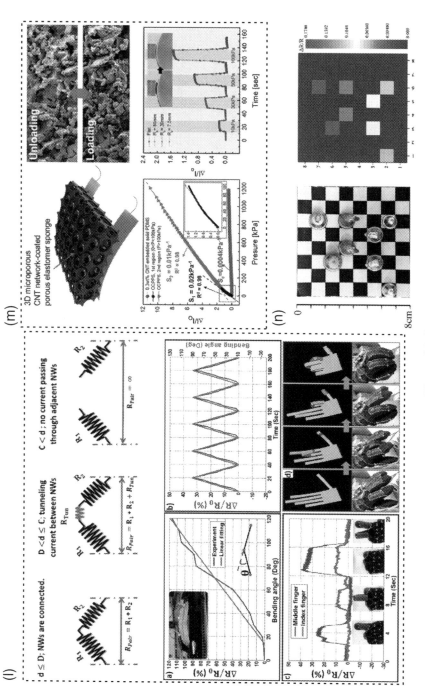

Figure 3.2 Cont.

antenna on an elastomeric substrate has been developed with a blade cutting tool and a series of transfer processes, as depicted in Figure 3.2c. The copper tape was attached to the blade cutter mat through a water soluble sacrificial layer and then stretchable serpentine-shaped geometry is realised on copper tape after removing the unwanted portion using a blade cutter (Figure 3.2d). Then, the serpentine copper tape antenna structure is transferred to dry-photoresist (PR) blue tape (Figure 3.2e) and then transferred to a semi-cured PDMS layer on a wafer (Figure 3.2f). After dissolving the dry-PR blue tape in acetone, a stretchable antenna structure is achieved. The fabricated stretchable antenna demonstrated an inductance value of 1.89 µH [97]. Such antennae are useful for transmitting continuous sensor data wirelessly for wearable real-time monitoring applications.

3.2.2 Metallic Nanomaterials

The development of metal-based stretchable electrodes has been achieved by careful geometrical engineering providing structural stability during the mechanical deformation. Such structures require careful fabrication methodology and engineering techniques, posing a significant barrier in the adaptation of techniques for large-area electronics applications. In this regard, one-dimensional (1D) nanomaterials such as metallic NWs and CNT-based network/thin-films have been actively explored. These materials can be deposited via low cost, large-area electronics compatible techniques [169], such as drop-casting [170], inkjet printing [171, 172], spray coating [173, 174], spin-coating [137] and screen printing [175, 176]. The aforementioned deposition techniques result in the network of 1D material, resulting in a percolation pathway offering highly stretchable conductive electrodes. The electrical and mechanical properties of the stretchable electrodes based on the 1D materials are governed by various factors such as the dimension (length and diameter) of the 1D materials, density of the material, junction resistance arising from the interaction between materials, adhesion between the 1D material, and substrate and network behaviour under mechanical strain. Generally, higher dimensions and density result in superior performance owing to a higher probability of percolation pathway.

Ag NWs are one of the popular nanostructures, owing to their high work function (4.5 eV) and ability to form a network with low sheet resistance (10–20 $\Omega/\square$) at 80% optical transmittance [177]. Such properties have also led to the investigation of Ag NWs as a leading candidate for transparent electrode applications. The electrical performance of NW-based electrodes is affected by factors such as junction resistance and poor adhesion between the NWs and substrate. Therefore, several studies have been carried out to investigate the role and to improve the electrical performance of NW electrodes. An example of homogenous distribution of spray-coated Ag NWs obtained by modifying the PDMS surface with polydopamine is shown in Figure 3.2g [178]. The polydopamine improves the hydrophilicity of the PDMS surface to ensure the homogenous distribution of Ag NWs and this eventually results in a stretchable electrode with 80% transmittance and sheet resistance of ~35 $\Omega/\square$, which remains unchanged under a 15% elongation. Junction resistance arising from an inter-NW interaction is another major issue affecting the electrical performance of NW-based electrodes. The junction resistance within the NW network can be minimised by soldering NWs together. Two major strategies adopted in this regard are: (1) development of hybrid material comprising NWs and other materials (Figure 3.2h); and (2) high-temperature thermal annealing treatment of NWs [179, 180]. In the former approach, materials such as GO [83], CNT [181] and graphene [137, 182] have been explored to improve the junction resistance of the NW-based network. For example, in a GO–Ag NW hybrid material combination, the GO wrapped around the NW structure aids in the reduction of junction resistance between the NWs [83]. This hybrid structure exhibited a very low sheet resistance of 14 $\Omega/\square$ and 88% transmittance at 550nm. Further, it also exhibits high stability, with less significant variation in the sheet resistance (2–3%) over a 12,000-cycle, showing its potential in stretchable electrode applications. The high-temperature thermal annealing of NW networks has also been shown to improve the junction resistance between the NWs owing to the fusing of NWs, as shown in Figure 3.2i [183, 184]. Besides the aforementioned approaches, other techniques such as directional alignment of NWs with solution process techniques have also been reported in the improvement of electrical and mechanical properties of a NW-based electrode. For example, a cross junction of Ag NW networks can be formed by coating Ag NWs by a meniscus-dragging process onto a substrate pre-coated with Ag NWs, rotated by 90° [185]. This results in a cross-junction of Ag NW film, as shown in Figure 3.2j, with a sheet resistance as of 14 $\Omega/\square$ and optical transmittance of 85%.

Copper NWs are another popular metallic nanostructures that have been widely studied for the development of stretchable electrodes [190, 191]. They have been studied as an alternative to Ag NWs owing to their low cost. Further, their electrical, optical and mechanical characteristics are similar to that of Ag NWs. However, Cu NWs are easily oxidised – as a result, their electrical properties are affected. It is critical to address this issue for Cu NWs for them to become a viable candidate for conductive electrodes. The approaches used to address the oxidation of Cu NWs include coating of thin Ni over Cu NWs (Figure 3.2k) or embedding the NWs within a polymer matrix, and so on [186,

191, 192]. These approaches help to maintain the stable electrical performance over time. These NWs can also be embedded in soft and elastomeric materials for various sensor applications. Figure 3.2l depicts the interaction of NWs inside the PDMS matrix with three different conditions: (1) ohmic contact when the centre lines of adjacent NWs are at a distance less than the diameter of the NW itself; (2) tunnelling effect when adjacent NWs are in a distance greater than their diameter and less than the cut-off range; and (3) no connection when the distance of adjacent NWs are greater than cut-off range [187]. Based on these conditions, the NW–elastomer composite-based piezo resistive materials have been used for detecting physical stimuli. For example, the composite conducting film made from Ag NWs embedded in PDMS has been used as a strain sensing material to detect the motion of fingers (Figure 3.2l) [187]. The piezoresistive strain sensors demonstrated tunable gauge factors in the ranges of 2 to 14 with a 70% stretchability useful for human–machine interface application.

3.2.3 Carbon-Based Conductors

Graphene and CNT are two most of the most popular carbon-based allotopes widely studied in the development of flexible and stretchable electronics. This is primarily attributed to their unique electronic, mechanical and optical properties that have led to exploring them for various applications both as semiconductors and conductors [193]. Like inorganic–elastomer composites, the composite materials using CNTs–elastomer have been widely explored for sensing applications. Figure 3.2m shows the usage of CNT-incorporated porous elastomer composite in 3D architecture for pressure sensing application [188]. The sensor responded to a wide pressure range, varying from 10 Pa to 1.2 MPa, with a linear sensitivity of 0.01–0.02 kPa^{-1}. Likewise, incorporation of iron particles inside the CNT–PDMS composite revealed bimode sensitivity to pressure and magnetic field [189]. For demonstration, the sensor was attached to the chess board and two different dimensional magnets were attached to the bottom of silver- and gold-coloured chess pieces. As shown in Figure 3.2n, the mapping data provided the details regarding the location and the colour of the chess coins using the pressure and magnetic bimode sensors. In this section, we will primarily focus on their application as electrodes.

Carbon nanotubes have been popular for the development of stretchable electrodes owing to their electrical and optical properties. The latter also makes them popular for the development of transparent stretchable electrodes. The highly entangled nature of CNTs within its network aids their ability to withstand mechanical deformation, allowing multiple electrical paths that are useful for maintaining a good electrical property while stretching. The simple approach is the development of a randomly oriented CNT network on an elastomer substrate through the spray-coating technique [194]. A spring-like CNT structure realised on stretched PDMS can withstand up to 150% strain, while exhibiting a conductivity of 2,200 S/cm in the stretched state. These electrodes have also been utilised to develop an array of capacitive pressure and strain sensors [194]. Direct synthesis of CNT film formed by floating catalyst chemical vapour deposition (CVD) has been demonstrated to form a stretchable film with electrical conductivity of 2,000 S/cm [195]. The PDMS can be used to improve the fracture strain up to 60% compared to the 10% of CNT film alone [196]. Structural engineering such as buckling has also been employed for stretchable CNT electrodes. This can be achieved by transferring the CNT film on to pre-strained PDMS substrate. The CNT film coated with Au/palladium could improve the electrical conductivity and can accommodate up to 100% strain with small resistance change (4.1%) [197]. The CNT-based stretchable conductors using electrospun PU and dip-coating process have also been demonstrated [198] with PU nanoweb dip-coated in a CNT ink. These conductors exhibit a sheet resistance of 424 $\Omega/\square$ and stable electrical response at 100% tensile stain, thereby showing their promise as a stretchable electrode.

As mentioned earlier, graphene's unique electronic properties have led to its investigation as both a conductor and semiconductor [199–201]. In particular, the high transmittance of graphene has led to it being used for transparent electrodes [202–4] and transparent e-skins [193, 203, 205]. In particular, reduced GO (rGO), CVD-grown graphene has been explored for electrode applications. The CVD graphene that is grown on Ni substrate and transferred on pre-stretched PDMS (12%) has been shown to have a stable resistance of up to ~12% stretching and an order of magnitude change in resistance at ~25% stretching [206]. These unique electronic properties of graphene have also been explored in developing an all-graphene FET (GFET) [201, 207]. All printed stretchable GFET, shown in Figure 3.3e, comprises of a monolithically patterned graphene channel, the source and drain electrodes with ion-gel serving as the gate dielectric. The GFET demonstrated a stable operation under a 5% strain even after 1,000 cycles. Besides its use as an electrode, CVD graphene has also been studied for interconnect applications. For example, the serpentine-shaped graphene was employed as a stretchable interconnect with inorganic LEDs [208]. The electrical characterisation of the structure revealed it was capable of handling up to 100% strain. Additionally, CVD-based graphene woven fabric-based structures have been explored in the development of stretchable electronic applications [209].

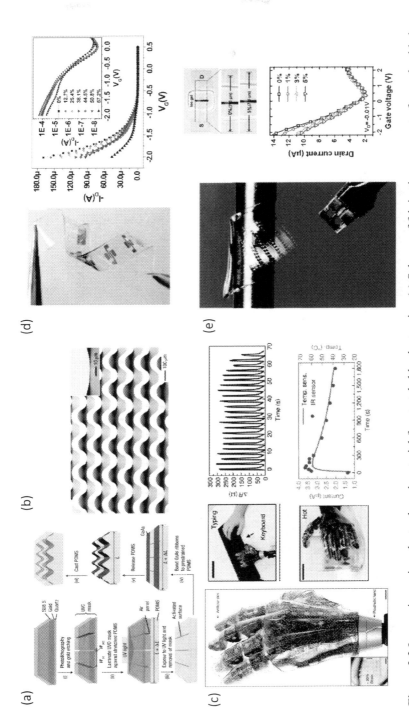

Figure 3.3 Inorganic semiconductor-based material for stretchable electronics. (a) Schema of fabrication process to form a buckled monocrystalline inorganic semiconductor structure. (b) Formation of buckled GaAs nanoribbon structure. Reprinted with permission from

Caption for Figure 3.3 (cont.)

[260]. Copyright (2006) Nature Publishing Group. (c) Photograph of stretchable e-skins on prosthetics integrated with silicon nanoribbon-based temperature, strain and pressure sensors. Scale bar: 1 cm. Image of the prosthetic tapping on the keyboard and holding a cup of hot water. The electrical response of the corresponding pressure and temperature sensors are shown on the side. Reprinted with permission from [261]. Copyright (2014) Nature Publishing Group. (d) Stretchable CNT-based FET on a PDMS substrate and its electrical characteristics up to a strain of 57.2%. Reprinted with permission from [262]. Copyright (2011) American Chemical Society. (e) Stretchable ion-gel-gated GFET on PDMS substrate. Microscopic image of stretched GFET up to 5% and its corresponding transfer characteristics. Reprinted with permission from [207]. Copyright (2014) American Chemical Society.

3.2.4 Organic Conductors

Organic conductors have emerged as a popular candidate for stretchable conductors owing to their mechanical softness and printing compatibility, making them the ideal choice for the development of large-area electronics. Some of the organic conductive polymers studied for stretchable electrode applications include polypyrrole [210], poly (3,4-ethylenedioxythiophene) polystyrene suldonate (PEDOT:PSS) [211] and polyaniline (PANI) [46]. Table 3.2 provides an overview of various organic conductors-based stretchable electrodes.

Among the aforementioned polymers, PEDOT:PSS has been extensively studied for various applications owing to its electrical conductivity (~100 S/cm), mechanical softness (stretchable up to 10%) [134, 211, 215] and transparency in the visible range [216, 217]. Despite its popularity, PEDOT: PSS has drawbacks such as modest electrical conductivity, stretchability and poor adhesion, especially to hydrophobic polymeric surfaces. These drawbacks can be addressed by doping PEDOT:PSS by introducing various surfactants such as dimethyl sulfoxide (DMSO) [23, 218], zonyl [211] and Triton X-100 [214]. The DC electrical conductivity of PEDOT:PSS increases from ~0.8 S/cm

Table 3.2 Summary of organic conductors-based stretchable electrodes

Material	Conductivity	Reliability	Stretchability	Ref.
PANI	>100 S/cm	N/A	0.5–300%	[46]
Polypyrrole with p-toluenesulfonate	171 Ω cm^{-2}	2,000	30%	[210]
PEDOT:PSS with zonyl fluorosurfactant	260 $\Omega/\square$	1,000 (30% strain)	~188%	[211]
PEDOT:PSS with zonyl fluorosurfactant	46 $\Omega/\square$	5,000 cycles (0–10%)	10%	[212]
PEDOT:PSS with stretchability and electrical conductivity enhancer	4,100 S/cm	1,000 cycles (100%)	800%	[213]
PEDOT:PSS nanofibril with Trinton X-100	250 $\Omega/\square$	500 cycle (ε = 10.3%)	ε = 14.3%	[214]

to 80 S/cm with the addition of DMSO [219] and to ~1,000 S/cm with the use of surfactants such as ethylene glycol and DMSO [217, 220]. Similarly, the use of zonyl has been shown to improve the mechanical resilience of PEDOT:PSS film, along with an increase in electrical resistance of the film by a factor of 2 under 50% strain while still exhibiting a significant conductivity even under 188% strain [211]. The durability of such a structure can be further improved by the formation of buckled film deposited on a pre-stretched elastomer. Such structures have been shown to have stable electrical conductivity over 5,000 cycles for 0–10% strain, thereby demonstrating the suitability as stretchable electrodes in applications such as displays, solar cells and so on [212]. Another challenge associated with PEDOT:PSS is its uniform coating on a hydrophobic surface. It has been demonstrated that the addition of non-ionic surfactants such as Triton X-100 to PEDOT:PSS results in the reduction of the surface energy of PEDOT: PSS, thereby improving its wettability of its hydrophobic surface. In addition, it also results in the formation of PEDOT nanofibrils with enhanced electrical conductivity (830 S/cm) while also exhibiting a 96% transmittance at 550nm. The nanofibrils also enable an electrical stability at 10.3% strain [214].

3.2.5 Nanocomposites

The conductive nanocomposites are studied as an alternative to the aforementioned materials for the development of stretchable conductors. The conductive nanocomposites comprise a network of conductive nanofillers within the elastomeric polymer matrix [221]. Some of the conductive nanofillers investigated for nanocomposites include carbon nanomaterials (e.g., carbon black, CNT and graphene) and metallic nanomaterials (e.g., NPs and NWs of the Au, Ag, Cu, etc.). The minimum volume fraction of conductive nanofillers required to attain the conductive electrical path within the polymer matrix is the referred percolation threshold [222]. In general, to achieve the required electrical conductivity, a higher volume fraction of conductive nanofillers exceeding the percolation threshold is used during the synthesis of the nanocomposite. However, high percolation often results in the degradation of the mechanical and chemical properties of the polymer matrix. Further, the higher volume fraction of the conductive fillers would also increase the cost of manufacturing, which is not desired when it comes to large-scale manufacturing. In practice, factors such as aspect ratio and the dispersion of the conductive fillers within the polymers matrix also affect the percolation threshold. In particular, high aspect ratio fillers can effectively reduce the percolation threshold as predicted through simulations [223]. A second approach to reduce the percolation threshold is to ensure homogenous distribution of the conductive fillers within the polymer matrix. In this regard, various techniques such as ball milling, ultra-sonication, grinder milling and jet-milling have been employed to achieve a homogenous dispersion. Among these, jetting-milling is favoured with respect to the other three, which affect the aspect ratio of the filler, for instance, shortening the CNT. Similar to these materials, various geometrical structural engineering techniques such as serpentine-, coil- and net-shaped structures have also been employed to enhance the stretchability of the composite structures.

3.2.6 Liquid Metals

Liquid metals have been widely investigated in the development of stretchable electronics and have been employed as stretchable interconnects, antennae, wires and functional materials [224]. It has been of particular interest due to its ability to withstand a large deformation, which in theory is limited only by the mechanical properties of the encapsulating material. Liquid metals include mercury, gallium and its alloys. The low toxicity and no vapour pressure of gallium and its alloys has made it a popular candidate [225]. Some of the popular gallium-based liquid metal alloys include eutectic gallium indium (EGaIn) and Galinstan (an alloy of gallium, indium and tin) [226–8]. Liquid metal alloys can be patterned with techniques such as inkjet, stencil lithography, microfludic, lift-off and laser ablation to pattern metal films on an elastomer [51, 229–33]. Further, as the liquid metals are easily oxidised, they can be 3D printed both out of plane. As mentioned, the maximum strain that the liquid metal-based structures can withstand is predominantly determined by the encapsulation material, therefore, microstructuring of PDMS and thermoplastic materials have been explored to aid in the structures that could withstand larger deformation. For example, EGaIn enclosed into the core of the thermoplastic polymer, poly-SEBS fibres can withstand a strain up to 800% [234]. Liquid metals have also been explored as self-healing structures for conformable electronic applications [235].

3.3 Active Materials

The active channel materials, which changes its electrical or mechanical property under external stimuli, are one of the key building blocks in the development of stretchable electronics. In this aspect, semiconductor materials whose electronics properties lie between that of conductors and insulators have been widely used. Such materials have been explored in the development of the FETs, LEDs and solar cells [236, 237]. In pursuit of the development of devices, various active materials have been investigated and this section provides an overview of such materials.

3.3.1 Organic Semiconductors

Conjugated polymers have been the primary driver in the development of flexible electronics owing to their intrinsic flexible nature and electrical properties. Nevertheless, conjugated polymers exhibit only a limited stretchability and as a result, found only limited success in the development of the stretchable electronics. Despite poor electrical performance compared to their inorganic counterparts (CNT, graphene, NWs, etc.), the conjugated polymers have been a popular choice owing to their low cost and facile large-scale processability. Some of the organic semiconductors (OSCs) investigated in the development of the stretchable electronics include pentacene, P3HT and dinaphthothienothiophene (DNTT) [238, 239]. Approaches such as geometrical structural engineering, altering the molecular packaging of the polymer, and improving the adhesion between the polymer and substrates are a few techniques employed to improve the stretchability of conjugated polymers. Geometrical structural engineering includes techniques such as buckling of semiconductors and the use of stretchable interconnects to connect rigid components of the circuit [59, 136]. As an example, an organic field effect transistor (OFET) capable of withstanding a 12% strain has been developed on topography-rich PDMS substrate [59]. Similarly, the interconnect-based approach has been employed for the development of stretchable organic transistors and LEDs. This approach has enabled the development of displays which could be stretched by 30–50% without any noticeable change in its performance. However, such approaches are currently not suited for the development of a high-density integrated circuit, and this has led to the investigation of more intrinsically stretchable semiconductors. In this regard, a blend of self-assembled P3HT NWs and PDMS has been shown to be promising with the active layer exhibiting excellent stability up to 10% strain [240].

The second approach explored to enhance the stretchability of conjugated polymers involves tailoring the molecular packaging of the polymers. For example, a copolymer of polyethylene with P3HT can withstand stretching up to 600%, and an OFET device comprising of this as an active layer has demonstrated mobility of 2×10^{-2} cm^2/V·s and an I_{on}/I_{off} ratio of $\sim 10^5$ [241]. Similarly, other conjugate polymers such as DPP crosslinked with a PDMS oligomer have also been studied as an intrinsically stretchable polymer-based active layer [242]. The cross-linking with PDMS enhances the conjugated polymer's mechanical stability, with 150% stretchability achievable with no crack formation. Further, the OFET fabricated with the conjugated polymer exhibits mobility of 0.4 cm^2/V·s after 500 cycles at 20%. The use of cross-linking has also been explored to develop a healable and intrinsically stretchable semiconductor based on the same conjugated polymer with different ratios of 2,6 pyridine dicarboxyamide incorporated into it [243]. The OFETs developed with a cross-linked conjugated polymer semiconductor exhibit mobility of 1.12 cm^2/V·s at 100% strain along the direction perpendicular to the strain.

3.3.2 Inorganic Semiconductors

The monocrystalline inorganic semiconductors have underpinned several advances in electronics over the last six decades. Silicon has been the workhorse material for the development of memory devices, transistors and integrated circuits, while gallium arsenide (GaAs) and gallium nitride have been explored in optoelectronic and power electronic applications [244, 245]. However, the existing technology cannot be directly used for the development of flexible/ stretchable electronics owing to its limitations such as ability to realise devices on planar substrates and the high processing temperature which is incompatible with a polymeric substrate used in stretchable and flexible electronics [22, 246]. To address these challenges, novel techniques in terms of material, structural engineering and integration techniques have been developed. For example, thinning of bulk Si wafers has been demonstrated towards the development of flexible electronics [247–51]. However, further refinement of these approaches is needed for them to be adopted for stretchable electronics. The interest in the use of monocrystalline inorganic semiconductors for the development of the flexible/ stretchable electronics is attributed to their superior electrical performance compared to OSCs, thereby enabling the development of devices on par with the traditional wafer technology. This has led to the investigation of micro- and nanoscale structures of monocrystalline inorganic materials such as NWs and nanoribbon/membrane, and so on [16, 61, 250–2]. The synthesis of nanoscale structures can be broadly classified into bottom-up and top-down approaches [253–6]. In the bottom-up approach, the nanostructures are synthesised via a chemical synthesis process in a well-controlled ambient, resulting in precise dimensions of the nanostructures. In the top-down approach, the nanostructures are realised from a bulk/thin-film through lithography and subsequent etching process. For example, such strategies have been adopted to realise Si NWs and nanomembrane from bulk silicon or silicon on insulator wafers [257, 258].

The next critical step in adopting the inorganic material towards the development of flexible or stretchable electronics necessitates the transfer of the structure to a defined location with nanoscale precision. Over the years, various wet and dry transfer processes have been developed to achieve this. This includes techniques such as Langumir–Blodgett, dielectrophoresis, transfer printing and contact printing, and so on [257, 259]. Although the monocrystalline inorganic nanostructures are inherently mechanically flexible, further integration techniques are needed to adapt them for the stretchable electronics and to develop conformable electronics. For example, buckling of Si nanoribbons (see Section 2.1) has been adopted to develop stretchable Si nanoribbon-based transistors and p–n diodes [258]. The device showed a stable response up to −9.9% strain. Further, no significant change in the electrical response of the device was noted over ~100 cycles of compress, stretch and release. Buckling of Si nanoribbon can be achieved by transferring the nanoribbon to the pre-stretched elastomers, on releasing the elastomer, the compressive stress causes the nanoribbons to buckle and form a wavy structure (Figure 3.3a). The dimension of the wavy structure is controlled by the thickness of the nanoribbon. Further, control on the dimension of the wavy structure can be achieved by defining the adhesion site in the elastomer via lithography and UV/ozone treatment of both the substrate and the active area [260], resulting in an improved stretchability. The scheme of the fabrication process and the 'wavy' GaAs nanoribbon formed by this process is shown in Figures 3.3a and 3.3b, respectively. Further, the out-of-plane structures have been shown to realise stretchable Si nanoribbon-based circuits such as inverters [28]. The patterning of Si nanoribbons into various geometrical structures has also been explored to realise a stretchable Si nanoribbon-based e-skin for prothesis to detect stimuli such as pressure and temperature, and so on (Figure 3.3c) [261].

Besides Si nanoribbons, carbon-based nanomaterials such as CNT and graphene have also been investigated in the development of stretchable electronics. Besides their use as conductors, CNT and graphene have also been utilised as semiconductors in the development of electronic devices. This has led to the investigation of carbon allotrope-based channel material for the development of stretchable FETs, and so on, such as the one based on CNTs shown in Figure 3.3d [262]. FET comprises of a polyfluorene-wrapped semiconducting CNT channel and ion-gel as the gate dielectric. The fabricated FET is capable of withstanding a strain up to 50%, with this stain being limited due to the fracture of the ion-gel gate dielectric. The I_{on}/I_{off} of $>10^4$ and low operation voltage (<2V) means that the device is electrically and mechanically stable over repeated stretching cycles, making them attractive for large-area applications such as e-skins and stretchable displays, and so on. Further, GFET based on ion-gel has also been explored for graphene-based stretchable FET [207] (Figure 3.3e). Besides ion-gel, other dielectrics such as transferred wrinkled Al_2O_3 have also been explored for stretchable electronics to develop CNT network-based transparent FET with a graphene-CNT electrode [263]. The FET demonstrates a high carrier mobility of 40 $cm^2/V \cdot s$, I_{on}/I_{off} of $>10^5$ and is capable of withstanding 20% strain. Further, the device showed no significant change in its electrical response during cyclic measurement (1,000 cycles), thereby showing its promise for stretchable electronics applications.

4 Stretchable System Fabrication Methods

To meet the integration requirement on soft and stretchable substrate, some of the conventional fabrication methods have been modified and this has resulted in methods such as transfer and contact printing [11, 21]. These techniques involve transferring the metals on to soft PDMS substrate. Direct deposition on PDMS using direct writing, shadow masking, screen printing and a conventional photolithography process are also being utilised for stretchable systems. In this section, we present the use and limitations of conventional fabrication techniques for stretchable systems. Various micropatterning transfer techniques, which have been developed to address the challenges related to the use of conventional fabrication methods for stretchable components on elastomeric substrates, are also discussed.

4.1 Conventional Fabrication Methods and Their Limitations.

Conventional fabrication techniques, such as lithography, shadow masking and wet etching, and so on, are often utilised to fabricate various geometries on an elastomeric substrate for a stretchable system [264].

4.1.1 Photolithography

Photolithography is utilised as the patterning technique in which the PR layer is selectively exposed to UV light to create stretchable geometries on a polymer resist. Sequentially, metallisation and lift-off steps are carried out to obtain micropatterned geometries [51, 265–7]. As a first step, the PR layer is spin-coated, soft-baked and patterned using a lithography tool. Following this, the metal layer is deposited on the patterned resist using direct deposition techniques, namely sputtering or an electron beam (e-beam) evaporator, and then the lift-off is carried-out using acetone or another resist stripper. However, this conventional process step is not compatible with elastomeric substrates, such as PDMS, due to the difference in the coefficient of thermal expansion between the spin-coated PR and the elastomeric substrate [268]. For example, the PDMS substrate is highly sensitive to temperature, which leads to undesirable cracks on the PR layer during the pre-baking step. In addition, the adhesion of the metal or any hydrophilic material is poor due to the super-hydrophobic nature of the PDMS substrate. Such challenges can be overcome with methods such as using an epoxy-based SU-8 as a sacrificial mask. The cracking of the PR on PDMS can be addressed by controlled temperature ramping [269]. Although there are no thermal cracking issues, the adhesion of a crosslinked SU-8 sacrificial mask to PDMS is very high and this requires mechanical peel-off using tweezers or scalpel tape. Alternatively, polyacrylic acid (PAA) can be utilised as an intermediate sacrificial layer between PDMS and the epoxy resist [268]. However, the pattern distortion and microcrack formation are unavoidable due to the large strain on the epoxy resist due to the expansion of PDMS.

4.1.2 Wet Etching

Wet etching selectively removes the unmasked area through chemical reaction. This helps to overcome the problems, such as pattern distortion and cracking, experienced with lithography/lift-off steps to achieve patterned geometries, such as serpentine-patterned Au on the elastomeric substrate. As shown in Figure 4.1a, prior to the lithography process, the thin metal layer to be patterned is deposited on the elastomeric substrate after oxygen plasma treatment [108, 271]. Sequentially, the lithography steps are carried out on top of the metal layer to define the pattern geometry. The patterned PR acts as the etch mask. Following this step, wet etching and resist stripping are performed to obtain patterned stretchable metal geometries on an elastomeric substrate without any undesirable cracks, as shown in Figure 4.1b. However, wet etching could result in undercuts due to an isotropic chemical reaction, which restricts its potential usage of this method when it comes to scaling down the pattern geometries.

4.1.3 Direct Deposition: With and without Mask

Direct metal deposition on to the stretchable substrate using a mask-based printing technique, such as shadow mask, screen or stencil printing, is another alternative process (Figure 4.1c). This simple and cost-effective process involves direct metal deposition through the predefined shadow mask to achieve stretchable geometries. The mask consists of micropatterned aperture with dimensions greater than 50 μm [272]. This technique is widely utilised for various conductive materials. For example, a 16 mm long and 200 μm wide Au strip has been fabricated using a polyimide shadow mask [273]. Similarly, stretchable electrochemical sensors using PEDOT:PSS ink on Ecoflex substrate has been fabricated by stencil printing [274] and Ag NW-based wearable multifunctional sensors have been realised through screen printing [275]. Advancement in shadow mask is made by utilising flexible parylene-C as a shadow mask to enhance the resolution by placing it in curvilinear surface [276]. However, the pattern resolution below 50 μm is not reliable using shadow mask-based direct patterning. Compared to the mask-based approach, the mask-less approach such as inkjet printing has significant advantages in terms of the

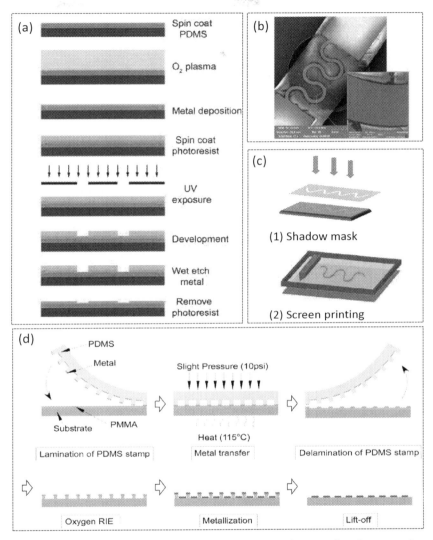

Figure 4.1 Fabrication of stretchable patterns on an elastomeric substrate using direct deposition techniques. (a) The schematic process flow for deposition and patterning of Au directly on the PDMS substrate using photolithography and wet etching. (b) An SEM image of serpentine-patterned Au on a PDMS substrate. (c)(1) represents the shadow-masking technique and (2) refers to the screen printing technique. Reproduced from [108]. (d) Metal transfer-assisted nanolithography technique on rigid and flexible substrates. Reproduced with permission from [270].

pattern resolution [277, 278]. This technique has the capability to deposit a variety of conductive inks such as Au, Ag and carbon-based conductive inks. However, the adhesion of conductive inks on hydrophobic elastomeric substrate is a critical issue. In addition, swelling and expansion of the elastomer substrate while thermal curing the conducting ink will result in undesirable cracks.m

4.1.4 Transfer Printing

Transfer printing has been developed as an alternative approach to address the limitations of direct metal deposition on an elastomeric substrate, such as cracking and delamination due to different coefficients of thermal expansion between the metal interconnects and the substrate. It utilises a patterned PDMS stamp, fabricated by soft-lithography, to transfer the patterned geometries onto unconventional substrates [279]. In this process, the target material is deposited on to a pre-patterned master stamp and transferred to the receiving substrate. The patterned PDMS master stamp is created using the conventional soft-lithography process. Sequentially, the master stamp is treated with the releasing layer to facilitate easy transfer before depositing the target material. During the transfer process, the master stamp is pressed against the final substrate under controlled pressure to selectively transfer the patterned geometry [280, 281]. Adhesion at the interface between the master stamp and the deposited material plays a crucial role in the transfer process. Poor adhesion is preferred at this stage. Accordingly, a less adhesive Au layer is often deposited directly on the master stamp, followed by the highly adhesive titanium (Ti) layer [282]. During the transfer process, a highly adhesive Ti layer is attached firmly to the final destination substrate and the less adhesive Au layer enables the easy transfer. With this technique, there is a limitation with respect to the metal thickness and transfer yield. To enhance the yield, a thin polymethylmethacrylate (PMMA) layer can be coated on the receiver substrate to enhance the adhesion at the receiver end. For example, chromium (Cr)/Au has been transferred to the unconventional substrate with PMMA coating using a metal transfer contact printing technique (Figure 4.1d) [270]. In this study, Cr/Au is utilised as a sacrificial lift-off layer to achieve an inverse replica of a master pattern.

4.1.5 Limitations of a Conventional Fabrication Process

The use of conventional fabrication processes is hindered by technical constraints related to integration of stretchable components on elastomeric substrates. For example, the thermal expansion coefficient (CTE) and the surface property of elastomeric substrates are vastly different from the typical silicon

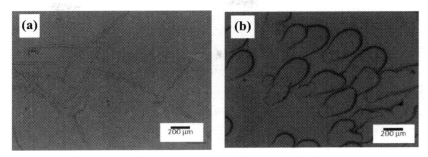

Figure 4.2 Challenges in performing conventional lithography steps on a PDMS substrate. An optical image of a spin-coated PR layer on a PDMS substrate, prior to lithography exposure steps, with (a) multiple cracks on the resist layer without using an intermediate layer and (b) formation of a scale-like undesirable structure with a polymeric interlayer. Reprinted from [283].

wafers used in the conventional fabrication techniques. Likewise, the direct metal deposition and other hard mask or stamp-based techniques result in low resolution. The fabrication of stretchable geometries on an elastomeric substrate using the conventional fabrication steps is also challenging due to the super hydrophobic nature of a soft PDMS surface, which leads to poor adhesion with metals and other organic material such as conducting polymers and PRs, and so on. Therefore, the established fabrication steps are not directly applicable, without modifications, for fabrication of stretchable interconnects on an elastomeric substrate. For example, using a conventional lithography process, the spin-coated PR layer such as SU-8 or AZ 5214E, and so on, on a PDMS substrate suffers from undesirable cracks during handling/annealing steps due to difference in the thermal expansion coefficient between PDMS and the resist layer (Figure 4.2a) [283]. To facilitate the lithography processing steps, the polymeric intermediate layer, including sodium polyacrylate, polystyrene or poly methyl methacrylate, was introduced between PDMS and the resist with the expectation of resolving the crack formation issue. However, the surface of the spin-coated resist layer on PDMS resulted in a non-uniform layer formation not suitable for lithography (as shown in Figure 4.2b) [283]. So far, the reported fabrication techniques have several limitations in terms of material adhesion, processing steps complexity, reproducibility, yield, pattern resolution and large-area fabrication. In this regard, scaling down the stretchable geometry down to a few microns resolution over a large area with a high yield are the contributing factors for realising stretchable systems. The following section presents the alternative fabrication strategy to achieve high-resolution patterned geometries on a PDMS substrate.

4.2 Fabrication of Stretchable Geometries on Elastomeric Substrates

To overcome the existing challenges in the fabrication process, either the pattern transfer technique (for high- performance inorganic devices) or printing techniques (for solution processable inks) are adopted. The pattern transfer technique is a promising approach to realise micro- and nano-resolution geometries on elastomeric substrate with high yield and reproducibility. In this approach, high-resolution patterns are first fabricated on a rigid substrate using a conventional approach and sequentially transferred to an elastomeric substrate. This pattern transfer is carried out in two ways, namely, a sacrificial layer-assisted transfer approach and a direct transfer approach. This section presents pattern transfer techniques assisted with – and without – a sacrificial layer, and printing techniques.

4.2.1 Sacrificial Layer-Assisted Pattern Transfer Technique

In the sacrificial layer-assisted transfer approach, high-resolution mesh/pattern geometries are fabricated on top of a sacrificial layer, namely potassium silicate (K_2SiO_3), aluminium (Al) or Ni, resting on a rigid substrate via a conventional lithography process and transferred to an elastomeric substrate [132, 283–5]. For example, metal nano-mesh has been fabricated on a silicon substrate using K_2SiO_3 as a sacrificial layer, and sequentially transferred to the elastomeric substrate after removing the sacrificial layer in the dilute hydrofluoric acid solution [132]. Similarly, a nanometre-scale conducting device can be fabricated on the Al sacrificial layer using e-beam lithography and successfully transferred to a PDMS substrate to implement a nanoresistor-based volatile organic compound (VOC) sensor [284].

Figure 4.3a depicts the schematic illustration of a step-by-step fabrication scheme of a high-resolution conductive Au pattern on a PDMS substrate using an Al sacrificial layer-assisted transfer approach [284]. The fabrication sequence begins with the deposition of a 150 nm thick Al sacrificial layer on top of a silicon substrate. Following this, PMMA as an e-beam resist is spin coated and ebeam lithography is performed to define the nanopatterns. Subsequently, metal deposition and lift-off has been carried out to obtain nanopatterned geometries on the Al/Si substrate. Prior to the transfer step, the samples are immersed in 3-mercaptopropyl trimethoxysilane (MPTMS) to enhance the adhesion between the metal and PDMS. Later, 175 μm thick PDMS is spin coated over the nanopatterns and cured. Finally, the Al sacrificial layer is etched using dilute hydrochloric acid (HCL) to successfully release the nanopattern-embedded PDMS substrate [284]. Figure 4.3b shows the

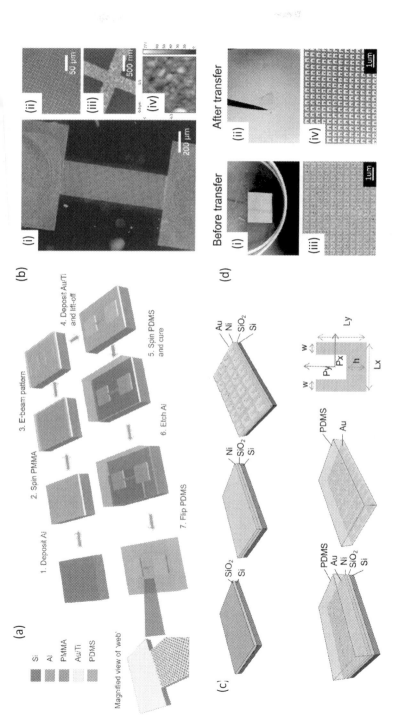

Figure 4.3 Sacrificial layer-assisted pattern transfer technique. (a) The schematic illustration of the fabrication process flow of an Au patterned conductive mesh structure transferred to PDMS via an Al sacrificial layer-assisted transfer approach, and (b) their microscopic images including (i) optical, (ii and iii) SEM and (iv) AFM images. Adapted with permission from [284]. (c) Schematic illustration of transferring metamaterials to PDMS via an Ni sacrificial layer-assisted transfer approach, and (d) their digital image, before ((i) photograph and (iii) SEM) and after ((ii) photograph and (iv) SEM) transfer. Adapted with permission from [285].

microscopic image of the high-resolution metal layer on a PDMS substrate. Likewise, Ni can also be utilised as the sacrificial layer to transfer nanopatterned geometries (Figure 3.9c) [285]. An example is shown in Figure 4.3d through the digital image of a micropatterned Au electrode on top of an Ni/SiO2/Si substrate (before transferring) and after transferring into the PDMS. However, large-area fabrication using the sacrificial layer-assisted transfer technique is quite challenging.

4.2.2 Direct Pattern Transfer Technique

Direct transfer of patterned metal geometries on to a PDMS substrate as an embedded structure without using any additional sacrificial layer has been explored as an alternative to the aforementioned methods [111]. This involves fabrication of a Cr mesh structure of micron-scale resolution on to indium tin oxide (ITO) substrate using photolithography. Sequentially, PDMS is poured, cured and peeled off from ITO to obtain a Cr mesh-embedded PDMS electrode [111]. The adhesion of Cr metal is weaker in ITO than in PDMS, which facilitates easy direct transfer. However, this transfer technique is not effective for any other metal patterns if there exists a good adhesion between metal and ITO substrate. Further, if there is a poor adhesion of patterned metal to the substrate, then obtaining a high-resolution pattern using lithography is difficult because of the pattern distortion during the lift-off process. Therefore, a new fabrication technique which can ensure both high resolution and good adhesion is essential. In this regard, a novel patterning technique using lithography and a transfer technique has been adopted to obtain honeycomb-patterned Au on an elastomeric substrate. Figure 3.10 shows the schematic flow for the fabrication of a honeycomb-patterned Au electrode on the ultra-thin PMMA substrate and its transfer technique to the elastomeric substrate [237].

4.2.2.1 Hexagonally Patterned Au Electrodes on Free-Standing PMMA

A 1.2 µm thick PMMA layer, spin coated on a rigid glass slide and annealed at 130°C for two hours is shown in Figure 4.4. Following a sequence of photolithography, metal deposition and lift-off, the hexagonally patterned Au electrode on PMMA can be obtained, as shown in Figure 3.10b. As the PMMA layer can get affected during the lift-off of PR in acetone, a 0.7 µm thick PVA layer can be spin coated as a sacrificial layer to make the metal lift-off process feasible on PMMA. Sequentially,

0.7 µm thick back anti-reflection coating (BARC) and 2.2 µm thick GXR positive PR is spin coated and annealed at 100°C and 115°C, respectively.

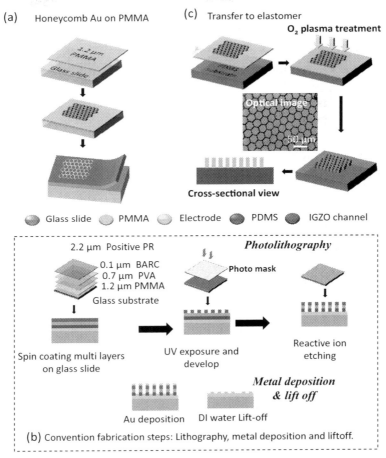

Figure 4.4 The schematic flow of (a) the preparation of a hexagonally patterned Au electrode on free-standing PMMA, (b) step-by-step photolithography, Au deposition and lift-off process and (c) the transfer of a patterned electrode onto the elastomeric PDMS substrate. Reprinted from [237].

As shown in Figure 4.4b, with UV exposure, developing and reactive ion etching (RIE), the hexagonal line trench on PVA/PR are obtained. Then, a 5/50 nm thick Ti/Au layer is deposited using an e-beam evaporator. The hexagonally patterned Au is obtained on PMMA after metal lift-off in deionised water. After fabrication of the hexagonally patterned Au, the PMMA substrate along with the pattern is readily peeled off the rigid glass slide by placing a tiny water drop at the interface between the glass slide and PMMA layer.

4.2.2.2 Transfer of Hexagonally Patterned Au onto the Elastomeric PDMS Substrate

Prior to the transfer of a patterned Au electrode, the PDMS substrate is treated under oxygen plasma using RIE to enhance the adhesion of PMMA with PDMS. As shown in Figure 4.4c, oxygen plasma treatment is also performed after placing the hexagonally patterned Au/PMMA on the PDMS substrate to remove the exposed PMMA region uncovered by the Au. The optical image in the middle of Figure 4.4c shows the hexagonally patterned Au transferred onto the elastomeric PDMS. Figure 4.5a shows the optical image of hexagonally pattern Au electrodes under different strains and the inset shows their respective magnified images. The geometry of the hexagonal pattern varies with increase in the strain and there are no visible cracks, even under the 1,000× magnification, at 20% strain. As shown in Figure 4.5b, dislocation of Au from the underlying PMMA is observed at higher strain. This means that the adhesion of PMMA to the elastomeric PDMS substrate is reasonably good for this transfer technique [237]. Further, the line width of the honeycomb geometry is around 2 µm and this technique has the capability to obtain high-resolution stretchable geometry over a large area. In addition, this process facilitates the transfer of honeycomb geometry on to the pre-strained PDMS substrate to further enhance the stretchability. Accordingly, the honeycomb geometry transferred on 30% pre-strained PDMS substrate (Figure 4.5c) demonstrates an enhanced stretchability of 75% strain with stable electrical performance (Figure 4.5d) [237].

4.2.3 Printing Techniques

Printing techniques have attracted significant attention for fabrication of complex geometries based on a wide range of materials through cost-effective adaptable fabrication techniques [286–8]. In the form of ink, the stretchable materials are printed on to elastomer substrate through a pre-designed screen as in screen printing technique, or directly written through opening of nozzles as in inject printing and the 3D stereolithography approach. The screen printing approach involves a screen with the desired geometrical structure and a squeegee to transfer the ink from one side of the screen to the other side under optimal force and pressure. The transferred inks directly printed on the elastomeric substrate, while moving the squeegee blade along the screen, leads to fabrication of a wide range of devices such as TFTs, antennae, sensors, fuel cells and electrode arrays [289–92]. For example, a highly stretchable electro chemical sensor and biofuel cell array was realised through a screen printing technique using CNT-based intrinsically stretchable inks [293]. Here, the combination of structural geometry and intrinsically stretchable materials is utilised to achieve two-degree stretchability.

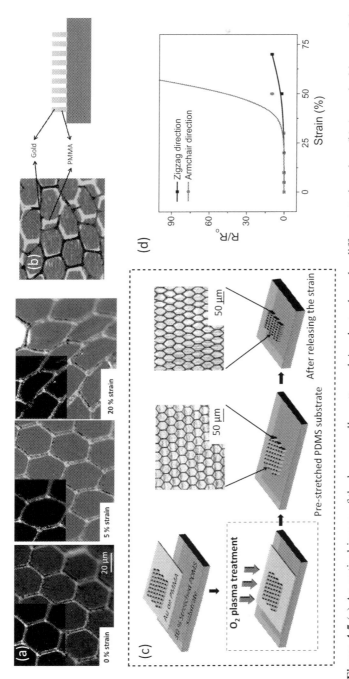

Figure 4.5 (a) An optical image of the hexagonally patterned Au electrode under different strain values. (b) An optical image of the hexagonally patterned Au electrode under 50% strain (left) and the cross-sectional schematic of the single-layer hexagonally patterned Au electrode on an elastomeric substrate. (c) A schematic drawing representing the transfer process of the honeycomb-patterned stretchable Au electrode on prestretched PDMS and their respective optical images, and (d) change in resistance under different strain. Reprinted from [237].

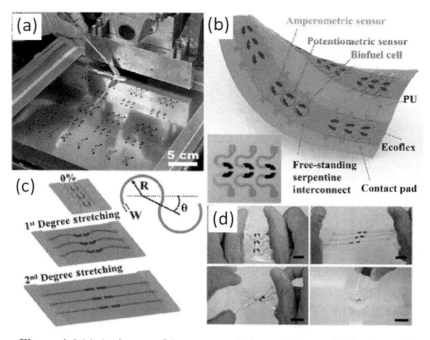

Figure 4.6 (a) An image of the screen printing technique. (b) A schematic illustration of electrochemical sensors and a biofuel cell array using serpentine interconnect with an intrinsically stretchable CNT electrode and (c) their two degrees of stretching induced by serpentine geometry and intrinsically stretchable CNT inks. (d) Photographs of the stretchable array under stretching, bending and twisting. Scale bar: 1 cm. Adapted with permission from [293]. Copyright (2016) American Chemical Society. (e) 3D printing of elastomeric metal-core silicone-Cu TENG fibres. Reprinted from [294]. Copyright (2020), with permission from Elsevier. (f) Photographs of the 3D-printed hollow 3D structure samples: a hyper small football, a lattice structure, a cellular sphere and its shape deformed with finger pressing; the mechanical and electrical response of the various structures under compressive strain. Adapted with permission from [295]. Copyright (2020) American Chemical Society.

The serpentine-shaped electrodes are screen printed using stretchable Ag/AgCl ink and, sequentially, an additive screen printing technique is used to print highly stretchable CNT ink on top of an Ag/AgCl electrode (Figure 4.6a). Following this, a series of printing processes was carried out to obtain an array of electrochemical sensors and biofuel cells, as shown in Figure 4.6b. The geometrically patterned intrinsically stretchable devices exhibit two-degrees of stretching upon

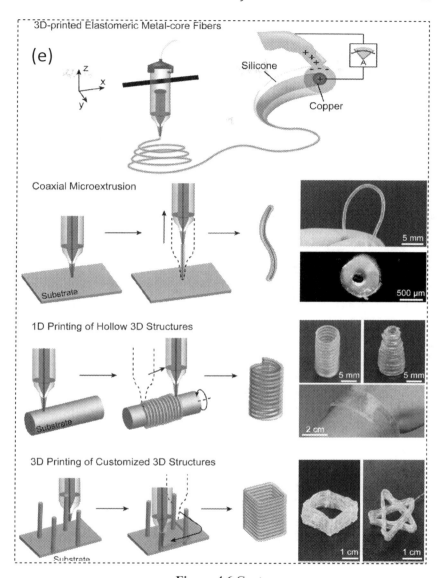

Figure 4.6 Cont.

external stress: (1) the geometrical pattern unwinds to external stress during first-degree stretching; and (2) the intrinsically stretchable composite ink made from a combination of CNT and elastomeric PU binder accommodates the second-degree stretching (Figure 4.6c and 4.6d). Accordingly, the fabricated devices demonstrated 500% stretchability. Similarly, organic stretchable inks are printed directly using an additive manufacturing technique, namely inject printing, to

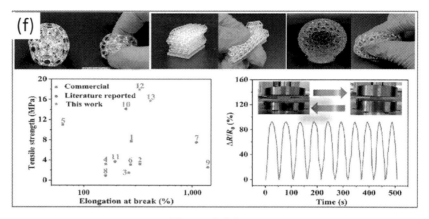

Figure 4.6 Cont.

achieve in-plane and out-of-plane 3D stretchable geometries. The stretchable inks are printed through a nozzle while the movement of a stage (holding substrate) is controlled by a pre-programmed pattern to achieve desired geometries. A few examples of 3D-printed geometries and their application in stretchable tribo-electric nanogenerator (TENG) devices and sensors are given in Figure 4.6e and 4.6f. The 3D printing of elastomeric metal-core TENG fibres using a multi-material microextrusion process was designed for a self-powered wearable device that generates current from proximity and touch [294]. The fabrication of elasto-meric metal-core TENG fibres with various geometries such as vertical, mem-branes, meshes, hollow 3D structures and customised 3D structures was demonstrated using planar, rotating and nonplanar substrates (Figure 4.6e). Likewise, a digital light processing 3D printer is used for printing complex structures. The mechanical properties such as tensile strength and elongation for 3D printed poly(tetrahydrofuran) units (PPTMGA-40) for sensor application was characterised. The strain sensor fabricated by coating an ionic hydrogel on the 3D surface demonstrated constant change in resistance under cyclic 30–80% compressive strain useful for soft robotics application [295].

5 Applications of Stretchable Systems

Stretchable systems have the potential to be conformably attach to human skin or any curvilinear surfaces for various applications including wearable health care monitoring, soft robotics, epidermal electronics, stretchable photo-volatiles and stretchable photodetectors [296–8]. Current stretchable systems often utilise rigid electronic devices with stretchable interconnects. In the previous sections, the materials and mechanism of stretchable systems were broadly discussed. In this section, the applications of stretchable systems in some of the areas such as health monitoring, energy devices and photodetectors are presented.

5.1 Wearable Electronics for Health Care Monitoring

There has been great progress in the development of stretchable and wearable sensors to enable non– invasive health monitoring techniques [299]. They are utilised for accurate and continuous monitoring of key human physiological parameters such as heart rate, blood pressure, skin temperature, electrocardiogram, electromyogram, electroencephalogram and respiration rate, and so on, using body fluids such as sweat, interstitial fluid, tears, saliva and urine, and so on [300, 301]. The wearable sensors on stretchable systems have the potential to conformably attach to the human body for real-time health monitoring and data collection. These methods have the advantage of saving the resources where the need for highly trained doctors or medical staff are not required. In this regard, stretchable sensing patchs for monitoring hear rate, pulse rate, oxygen level and sweat pH detection have been designed by coating pH-sensitive organic silicate on to commercial rigid blood oxygen sensors. An example of this type of sensors patch is the rigid blood oxygen sensors developed with a power circuit and processing circuit (for data storage and transfer) on a rigid platform with serpentine-shaped copper interconnects to enable stretchability [302]. The fabricated sensor patch demonstrated good stretchability, up to 35% strain, conformability and wireless capability for data transmission to monitor health condition in real time [302]. Similarly, wearable respiratory sensor has been fabricated by designing horseshoe-patterned Ag NPs using an inkjet printer to achieve a high accuracy of detecting the human respiratory rate [303]. Sweat in the human body contains abundant biochemical components useful for monitoring health. Wearable and skin-mounted electrochemical biosensors to detect glucose and pH in sweat have also been developed by coating CoWO4/CNT and PANI/CNT nanocomposite on to an Au stretchable electrode [304]. To enhance the stretchability and conformability, a fully stretchable nanoporous Au working electrode was fabricated on to a mogul-patterned PDMS substrate with curvilinear connected bumps and valleys to realise fully stretchable sweat glucose sensing patch [33].

Figure 5.1a shows the stretchable microfluidics-integrated sensing patch attached on a human hand. The liquid transport behaviour of the microfluidic-integrated sensor patch under various stretching conditions is shown in Figure 5.1b. A nanoporous Au active electrode (Figure 5.1c) for glucose sensing, an Au counter electrode and Ag/AgCl reference electrodes are fabricated on mogul-patterned

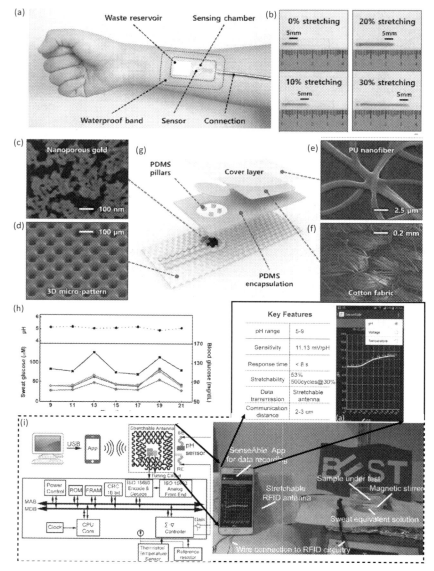

Figure 5.1 (a–g) A schematic of an omnidirectional stretchable/wearable microfluidics-integrated biosensor patch for blood glucose-level monitoring. The sensor is fabricated on a mogul-patterned PDMS substrate consisting of a bump and valley design. (a) The photograph of a fabricated patch conformably attached on to ahuman hand, and (b) an optical image of the sensor patch subjected to 0–30% stretching. SEM images of (c) a nanoporous Au (NPG) electrochemical electrode (top view), (d) 3D mogul-patterned PDMS substrate, (e) PU NFs and (f) stretchable cotton fabric. (g) Assembly of layered

Caption for Figure 5.1 (cont.)

components to achieve a fully stretchable microfluidics-integrated biosensor patch. (h) Monitoring by a sensor patch placed on human skin of pH (top) and glucose (bottom) in sweat after 20 min exercise. Adapted with permission from [33]. Copyright (2019) American Chemical Society. (i) System-level block diagram of stretchable pH sensor and a wireless pH data transmission unit using a serpentine-shaped RFID antenna (left) and the photograph representing the real-time pH monitoring and wireless data transmission for the sweat-equivalent solution (right) and the screen shot of smartphone App 'SenseAble' developed by Bendable Electronics and Sensing Technologies (BEST) group (top). Reprinted from [97]. Copyright (2018), with permission from Elsevier.

PDMS substrate (Figure 5.1d) to enable multidirectional stretchability. A microfluidic device with thin PDMS, PU NFs (Figure 5.1e) and cotton fabric (Figure 5.1f) has been integrated with the non-enzymatic nanoporous Au biosensor to form a microfluidics-integrated biosensor patch (Figure 5.1g). The mogul-patterned PDMS substrate enhances the omnidirectional stretchability with 30% extensive deformation [33]. The stretchable microfluidics-integrated biosensor patch along with portable potentiostat is attached on to a human arm to monitor sweat glucose levels, the sweat was generated by a 20 min cycling experiment. One-day monitoring of sweat glucose levels using the wearable sensor patch showed reliable performance (Figure 5.1h) even after harsh cycling motion. In this study, the data transmission is carried out through flexible wire connections. Another example for sweat pH analysis is the stretchable wireless system fabricated using serpentine-shaped interconnects with a stretchable RFID antenna [97]. In this case, the stretchable sensor patch has the ability to transfer data wirelessly through a stretchable RFID antenna. The stretchable pH sensor, with a graphite-PU composite-based pH sensing active electrode and an Ag/AgCl-based reference electrode, can withstand stretchability up to 53% strain (higher than the average strain of human skin (30%)) without compromising the electrical performance [305]. An example of the system-level block diagram of a stretchable wireless system for sweat pH monitoring is shown in Figure 5.1i. When coupled with a smartphone app, the stretchable RFID antenna allows the real-time data transfer without any external power supply. The geometry of the antenna is modified to a serpentine-shaped coil with an inner and outer radius of 1.5/2.5 mm with an 180° arc. The stretchable system using serpentine interconnects demonstrates stable performance at 30% for 500 cycles which leads to improved conformability of a pH sensor system.

5.2 Stretchable Electronics for Multi-functional Electronic Skins

The electronic devices, such as TFTs, strain sensors [306], pressure sensors [307], tactile sensors [308], photodetectors [309] and image sensors [310], have the ability to transform human activities into perceptible electrical signals to gain insight into the user activity level. Among them, sensors are used to detect and monitor external stimuli and transistor acts as a building block for signal processing and computation [311]. This section presents stretchable photodetector and functional devices (sensors and transistors) for soft robotics.

5.2.1 Stretchable Photodetectors

The stretchable photodetector attached to the nonplanar hemispherical surfaces (concave or convex) has a variety of innovative applications, such as mimicking the biological eye to aid the visual sensing in robots [312]. The organic–inorganic heterojunction materials, such as ZnO NWs on PEDOT:PSS or ZnO NWs on PVP fibres or semi-crystalline perovskite, have been utilised as active materials for photodetectors. This is owing to their superior properties such as high absorption coefficient, tunable optical properties, prolonged carrier lifetime and high flexibility [31, 256, 313, 314]. To achieve stable performance over long elongation, stretchable geometries including crumbled, wavy, kirigami and mogul-like patterns have been utilised [31]. For example, kirigami-patterned PDMS substrate has been utilised to fabricate Au@PVP NFs-based stretchable electrodes and a ZnO NWs@PVP NF-based stretchable channel for realising an ultrastretchable photodetector device. Figure 5.2a explains the fabrication and transfer of an Au@PVP NF electrode on a kirigami-patterned PDMS substrate. The device current-voltage characteristic (Figure 5.2biii) measured under dark and 365 nm UV illumination ($13.69 W/m^2$ intensity) with constant strain (0%) and bias voltage (5V) shows an increase in the current value of more than four orders of magnitude from 74 pA to 129 nA with a responsivity of 0.124 mA/W. A photodetector on a kirigami-patterned PDMS substrate demonstrates stable performance for up to 50 stretching cycles along the channel length under 80% strain (Figure 5.2b) [315]. Likewise, 3D micropatterned stretchable substrates with bump and valley architecture have been utilised to demonstrate 30% stretchability in a ZnO NWs@PEDOT:PSS-based photodetector (Figure 5.2c) [31]. A crumbled graphene active layer fabricated using pre-stretched PDMS substrate has also been utilised to demonstrate stretchable photodetectors [316, 317]. However, organic materials are not stable under atmospheric conditions [318] and therefore metal oxides such as V_2O_5 [319], ZnO [320], InGaZnO [321] and CdO [322] are utilised [323]. The photodetector based on metal oxide materials are eco-friendly and can bear the

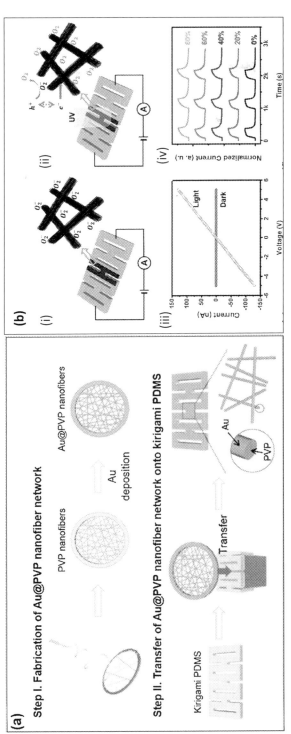

Figure 5.2 (a) Schematic representation of a Kirigami-based stretchable Au@PVP NF conductor network by a two-step fabrication process. First, (i) Au@PVP NF network fabrication through electrospinning a PVP NF network and Au deposition, and (ii) transfer of an Au@PVP NF network directly on to a kirigami-patterned PDMS substrate. (b) A stretchable photodetector on a kirigami-patterned PDMS substrate and its electrical characteristics. The UV light detection mechanism (i) in dark and (ii) under light illumination, and (iii) their respective IV characteristics; (iv) and comparison of UV detection performance under different stretching with strain varying from 0 to 80%. The stretchable photodetector on a kirigami-patterned PDMS substrate shows 80% uniaxial stretching without performance degradation. Reprinted from [315]. Copyright (2020), with permission from Elsevier. (c) A stretchable UV photodetector on a 3D-patterned PDMS

Caption for Figure 5.2 (cont.)

substrate and its electrical performance under stretching. (i) A schematic illustration of a stretchable photodetector fabricated on a mogul-patterned PDMS substrate, attached to a human hand for detecting the UV light; (ii) an SEM image of ZnO nanorods grown directly on to a 3D-patterned PEDOT:PSS layer resting on a mogul-patterned PDMS substrate; and (iii) performance of the stretchable UV detection under various stretching conditions varied for 0 to 30% strain. The device shows stable electrical response while subjected to various stretching conditions. Adapted with permission from [31]. Copyright (2017) American Chemical Society. (d) An intrinsically stretchable honeycomb-patterned InGaZnO-based inorganic photodetector and its sensing response under stretching. Schematic representation of a honeycomb-patterned photodetector stretched along (i) zigzag direction and (ii) armchair direction; (iii) their sensing performance under different stretching conditions varied form 0 to 20% strain; and (iv) their respective responsivity; the sensor shows stable sensing performance under 10% strain with pattern-induced stretchability and pattern-induced performance enhancement. Reprinted from [237].

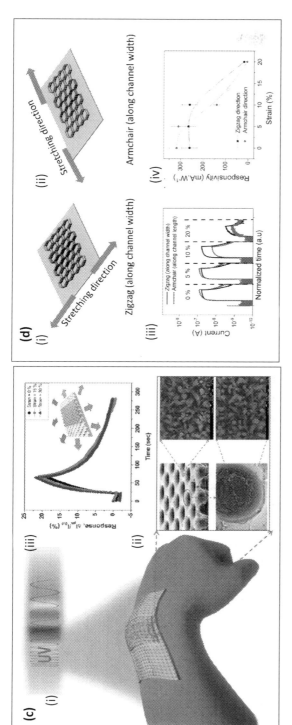

Figure 5.2 Cont.

harsh environment conditions but they cannot endure physical strain greater than 1% while being stretched, bent or twisted [16]. Nonetheless, the metal oxide NW-based devices are capable of handling greater strain, with the charge transport between individual NW networks attributed to a charge-hopping mechanism, which deteriorates the charge carrier mobility and the performance [324].

To achieve stretchability along with the good performance in metal oxide-based photodetectors, honeycomb-like patterning has been introduced with conducting metal and semiconducting inorganic materials. As an example, the honeycomb Au on PDMS demonstrates omnidirectional stretchability up to 20% (Figure 5.2d). Interestingly, owing to the increase in surface area of honeycomb geometry, the photodetector with a honeycomb-patterned IGZO semiconductor channel can exhibit 25 times higher responsivity than the plain (non-patterned) IGZO channel. Furthermore, the Au decoration on the honeycomb-patterned IGZO photodetector device reveals an excellent responsivity of 295.3 mAW^{-1}, which is 127-fold higher than plain IGZO, with an omnidirectional stretchability.

5.2.2 Stretchable Functional Devices for Soft Robotics

The e-skin capable of detecting various stimuli such as temperature, pressure, strain, force and sheer are necessary for robotics to mimic the functionality of human skin for effective interaction with humans and their surroundings [325]. Human skin has numerous receptors such as mechanoreceptors to feel the pressure, thermoreceptors to detect the temperature and nociceptors to detect pain. Likewise, e-skin capable of detecting and distinguishing various external stimuli is critical for soft robotics application. In this regard, stretchable strain sensors, pressure sensors, tactile sensors and temperature sensors have been widely explored. Several reviews exist that focus on soft e-skin [11, 20, 22, 326]. Under multi-modal stimuli, the sensor capable of responding to one selected stimulus and insensitive to the rest is highly needed. Considering this requirement, a robust stretchable pressure sensor array with Wheatstone bridge configuration to weaken the temperature effect and double-layer island configuration to remove the strain effect was demonstrated [327]. The sensor was subjected to multi-modal interference through an experimental demo: (1) sensors were placed on a robotic finger that then grasped a bottle containing water at two different temperatures (30–80°C, Figures 5.3a and 5.3b); (2) sensors were placed on a robotic finger by a dry and then wet human hand (Figure 5.3d); and (3) sensors were subjected to stretching while pressed with another hand (Figure 5.3e). The device

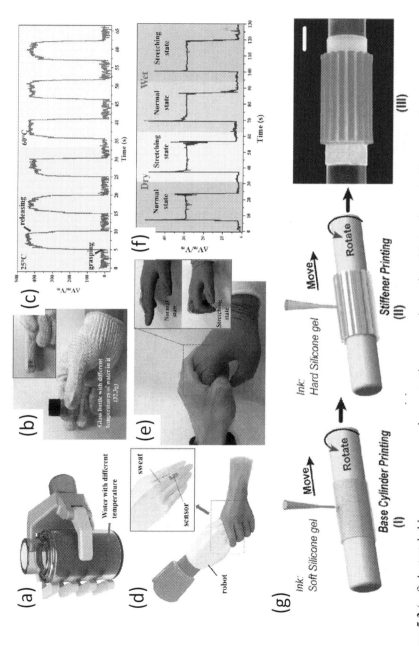

Figure 5.3 (a–f) A stretchable pressure sensor demonstrating a grasping action. (a) A schematic illustrating the sensor on a hand grasping a bottle of water; (b) an optical image of a sensor grasping a bottle; (c) pressure sensor characteristics under hot and cold condition; (d) sensor

Caption for Figure 5.3 (cont.)

grasping; (e) an optical image of a sensor grasping and under stretching; and (f) a pressure sensor response under different stretching and humidity conditions. Reprinted from [327]. Copyright (2020) with permission from Elsevier. (g–k) Artificial muscle printing and the displacement of an axis for lifting objects. (g) A schematic illustration of a direct ink writing process for fabricating artificial muscle; (h) geometry, simulation and an FEA model; (i) an equivalent elastic strain and actual experimental shapes of the artificial muscle; (j) axial contraction under different pressures obtained by an FEA model, analytical model and experiments; and (k) an optical photograph of using artificial muscles to lift objects. Scale bar: 13 mm. Reprinted from [328]. Copyright (2019) American Chemical Society. (l–n) A rubbery semiconductor elastomer composite-based strain sensor; (l) a photograph of sensors attached to the nitrile glove at a joint region; (m) a sensor response under the movement of the index finger; and (n) sensor mapping under the clenched fist motion. Reprinted from [329]. Copyright (2018) American Chemical Society.

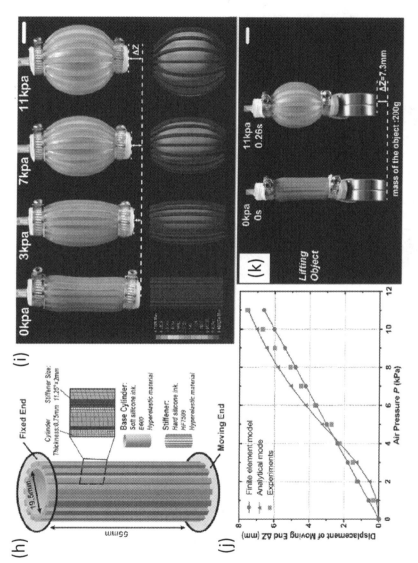

Figure 5.3 Cont.

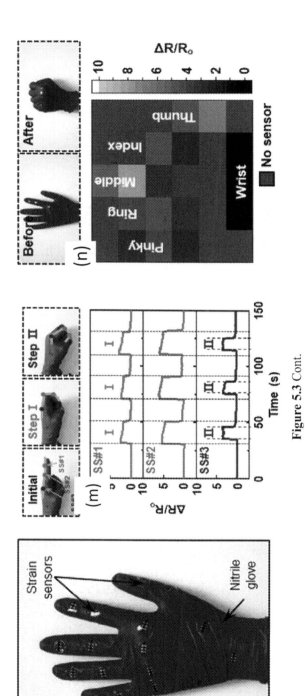

Figure 5.3 Cont.

revealed stable pressure sensing performance under multi-modal interface with the sensitivity of 2.6–14.05 kPa^{-1} under 0–74 kPa (Figures 5.3c and 5.3f) [327]. Likewise, a direct ink writing process was used to print artificial muscles using soft and hard silicone gel, as shown in Figure 5.3g [328]. A FEA model was used to understand the working principle of artificial bionic muscle by considering the radial stiffness and axial stiffness to achieve contraction of the muscle at different pressures (Figure 5.3h). The experimental and simulation results of artificial muscle contraction under different air pressure is given in Figure 5.3i and 5.3j. The axial contraction under air pressure is used to lift objects as shown in Figure 5.5k.

The direct ink writing process is also used to print silicone gel–conductive grease–silicone gel composite architecture for reliable strain sensor application [328]. For example, a strain sensor with rubbery semiconductors nanocomposite made of π–π stacked P3HT nanofibrils percolated in a silicone elastomer of PDMS has been reported [329]. The fabricated strain sensors demonstrated <12% hysteresis up to 100% strain with a linear response ($R^2 >$ 0.996) and high gauge factor of 32. Further, the strain sensor was attached to the finger joint region of the glove and the sensor response was characterised under the index finger's two-step bending motion and clenched fist motion to show the capability for mimicking the strain functionality of skin. Further, integration of sensors such as a piezoelectric pressure sensor with transistors reveals a 150 times greater sensitivity than a piezoelectric pressure sensor alone [333, 334]. For e-skin application, omnidirectional stretchability and conformability are the critical parameters. An omnidirectionally stretchable OFET was achieved by fabricating an omnidirectionally wrinkled alkylated DNTT derivative 2,9-di-decyl-dinaphtho-2,3-b:20,30;-f-thieno-3,2-b--thiophene (C10-DNTT) [330]. Figure 5.4a depicts an array of stretchable OFET using a parylene dielectric layer, C10-DNTT OSC, Al gate and Au source/drain. Figures 5.4b and 5.4c show the photomicrograph of a wrinkled OFET array with no strain and under different mechanical deformations. The stretchability of the OFET was demonstrated by attaching them on to the finger joints and the transistors mobility was characterised under a bending motion (Figures 5.4d and 5.4e). The OFET exhibited stable mobility under different mechanical deformations. Further, the transistor was attached to the balloon to introduce omnidirectional stretching (Figures 5.4f and 5.4g), which resulted in stable transistor performance. Likewise, a SOTT was realised using Ag NWs as source/drain electrodes, P3HT-PDMS blends as an organic semiconducting layer, and PDMS as a dielectric layer without gate electrode for tactile sensing application (Figures 5.4h and 5.4i) [331]. A touch or gentle tapping on top of a gate dielectric layer modulates the

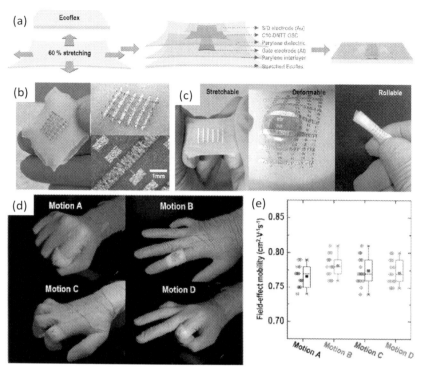

Figure 5.4 (a–g) Omnidirectional stretchable OFET array. (a) A schematic representing the OFET fabrication; a photomicrograph of the OFET array (b) under normal conditions and (c) under mechanical deformation; (d) a photomicrograph of the OFET array attached on to a finger joint under different bending motions and (e) mobility; (f)a photomicrograph of an OFET array attached to a balloon under normal and expansion states, and (g) their transfer characteristics. Reprinted from [330]. Copyright (2020) American Chemical Society. (h–j) A stretchable organic tribotronic transistor (SOTT) and their tactile sensing characteristics under stretching. Reprinted from [331]. Copyright (2020) with permission from AAAS under CC-By-4.0. (k) Illustration of chameleon-inspired stretchable e-skin. Reprinted from [332] under CC-By-4.0.

source-drain current. As shown in Figure 5.4j, the SOTT attached on to the finger joint demonstrated stable tactile sensing behaviour under a pristine and stretching state. Likewise, integrating the electrochromic polymer layer along with the pyramid-shaped PDMS/SWNT-sensing layer exhibits colour change under tactile touch, like a chameleon for wearable e-skin application (Figure 5.4k) [332].

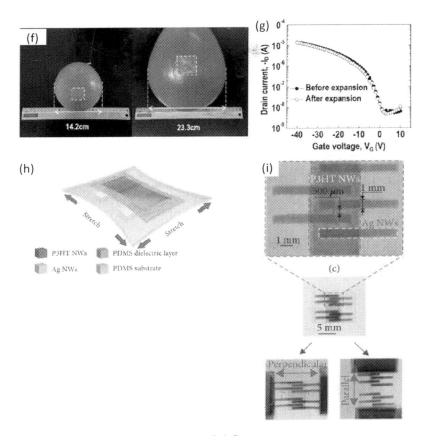

Figure 5.4 Cont.

5.3 Stretchable Energy Devices

An internally generated power source is needed for energy autonomous stretchable systems to drive various physical (pressure, temperature and strain) and chemical (pH, glucose, etc.) sensors. Therefore, mechanically robust, light-weight, low-cost and highly efficient energy devices such as supercapacitors, solar cells and batteries have been explored [23–6]. For stretchable batteries, all of the layers including electrodes (anode and cathode), electrolyte and separator should be able to endure mechanical deformation. The examples in this regard include the micro-honeycomb structures of graphene-CNT/ active material composite electrode [335]. As shown in Figure 5.5a, a micro-honeycomb-patterned active material entangled with graphene/CNT, fabricated using a silicon mould through a freeze-drying method is encapsulated with a physically cross-linked gel electrolyte as a separator. The fully assembled battery demonstrates omnidirectional stretchability with superior

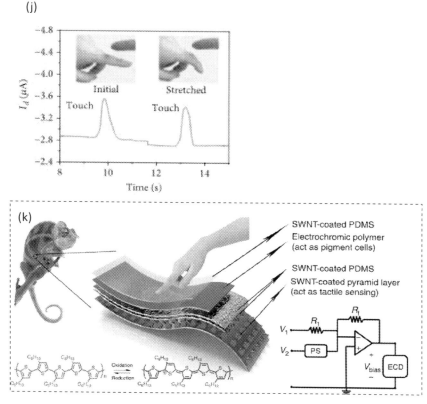

Figure 5.4 Cont.

electrochemical performance up to 50% strain with long-term stability up to 500 stretch-release cycles [335]. Likewise, several stretchable lithium ion batteries have been developed using different stretchable structures such as 3D porous sponge, wavy, spiral, mesh, kirigami and serpentine, and so on, to induce structural stability over 150% deformation [336– 41]. Similar to the batteries, the electrode with different structural geometries such as serpentine, wrinkle, kirigami and honeycomb structures have also been implemented to achieve a stable performance in super-capacitors [296, 342–5]. Unlike batteries, supercapacitors have the advantage of fast charging (nearly 1,000× faster than batteries), a long lifespan, fast transient response, high power density and wide operation temperatures, making them reliable and a promising alternative energy storage device for stretchable systems [346]. In general, stretchable batteries or supercapacitors with 2D-stacked architec-ture are not suitable for stretchable systems owing to the interfacial mismatch between electrodes, electrolyte and separator, which leads to some degradation

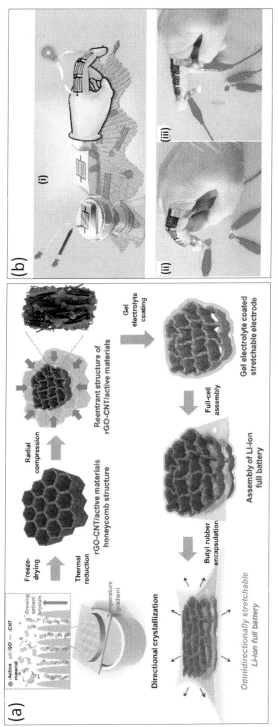

Figure 5.5 Stretchable energy storage devices. (a) A schematic illustration of the fabrication process flow of a stretchable lithium-ion battery using inwardly protruding micro-honeycomb-shaped graphene-CNT/active material geometry; the process flow includes directional freezing self-assembly, radial compression, a stretchable gel electrolyte and butyl rubber encapsulation. Adapted with permission from [335]. Copyright (2020) American Chemical Society. (b) Stretchable tandem planar MSCs fabrication and its wearable application: (i) a schematic illustration of a stretchable supercapacitor with wrinkled GCP film, achieved by dry transfer of GCP film on to prestretched elastomer substrate, and serial integration of GCP-MSCs on rubber glove; (ii and iii) the application of wearable/stretchable MSCs on a rubber glove to power the LED under different strain states. Reprinted from [347]. Copyright (2018), with permission from Elsevier. (c) A fully stretchable system with a strain sensor integrated with an array of MSC and solar cells using serpentine-shaped interconnects: (i)

Caption for Figure 5.5 (cont.)

a schematic illustration of a biaxially stretchable MSC array integrated with a solar cell and strain sensor using serpentine-shaped interconnects; (ii) a photographic image of a fully stretchable system conformably attached to the human hand to detect pulse and motion; and (iii) a comparison of a fully integrated system performance with a solar simulator and external power source. Reprinted from [349].

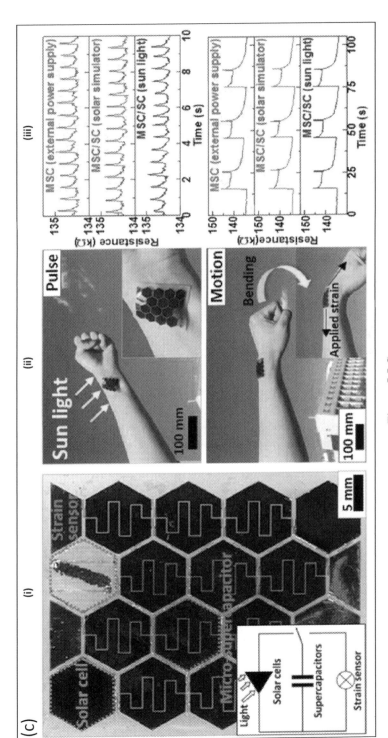

Figure 5.5 Cont.

effect during deformation. In this regard, a coplanar stretchable supercapacitor, which consists of two electrodes on the same plane separated by a small space filled with the electrolyte material, are better. For example, micro-supercapacitors (MSCs) with coplanar electrode configuration and interdigitate structure have been developed using graphene/CNT/cross-linked PH1000 (GCP) film on wrinkled substrate [347]. Figure 5.5b shows the fabrication of stretchable tandem GCP-MSCs using a customised interdigitated mask and dry transfer on to the pre-stretched rubber to achieve wrinkled architecture [347]. The resultant wrinkled GCP-MSCs demonstrate stable performance up to 200% stain under 8,000 repeated strain and release cycles. Figures 5.5b2 and 5.5b3 show the working of GCP-MSCs fabricated on a glove to power the LEDs under different stretching movements of a human finger. However, it is harder to integrate a wrinkled supercapacitor to fully stretchable systems. To address the issues related to the integration, planar MSCs and the gas sensor have been fabricated on a stiff platform of SU-8 with serpentine-shaped stretchable interconnects [348]. Similarly, a fully stretchable system with a strain sensor device with integrated energy harvesting and storage devices has been developed to self-charge and drive the sensors [349]. The interconnection of various devices such as a strain sensor, MSCs and solar cells through serpentine-shaped interconnects is shown in Figure 5.5c(i). The photographic image of the stretchable system shows serpentine interconnects connecting the active strain sensor device powered by an array of high-performance solid state MSCs, which are charged using commercial Si-based solar cells. The system when attached to the human skin provides a real-time pulse signal of radial artery and wrist motion, as shown in Figure 5.5c(ii). The MSCs in this case can be charged in different ways, such as using (1) an external power supply; (2) a solar simulator; or (3) sunlight (Figure 5.5c(iii)). The integrated strain sensors demonstrate a pulse rate of 15 beats per 10 seconds, which is similar to the radial artery pulse signal. When connected to wrist, the sensor can respond to the wrist bending motion through varying resistance values (134 Ω (straight) to 145 Ω (under bending state) and then back to 134 Ω (at straight state)). The fully stretchable system with a strain sensor integrated with MSCs and solar cells using serpentine-shaped interconnects demonstrates stable performance over 30% stretching strain, which is higher than the stretchability of human skin.

6 Conclusion

The development of stretchable electronics has gained significant pace in recent times. This is driven by potential novel applications in the biomedical field, consumer electronics and e-skins, and so on, where devices are expected to operate under mechanical deformation. These requirements have also fuelled the studies towards the development of novel materials and new engineering and fabrication technologies to advance the field of stretchable electronics. This Element has presented a comprehensive analysis of geometrical design and materials for stretchable systems along with the recent progress from the point of view of applications. In pursuit of the realisation of stretchable electronics, tremendous efforts have focused on the intrinsically stretchable components such as electrodes, interconnects, conductors and semiconductors, and so on. Such components have been developed using novel materials such as organic materials, NWs and carbon-based nanomaterials such as CNTs and graphene. Another approach involves the employment of various geometrical structures to improve/instil stretchability. This approach has proved to be successful in the development of monocrystalline inorganic material-based stretchable devices. The latter approach involves the transfer of suitable materials on pre-stretched elastomers, which forms a buckled structure on the release of the strain and thus results in improved stretchability. These two approaches have their pros and cons in the development of stretchable electronics, which are also discussed in this Element. While the former approach enables the larger stretchability with limited stability, the geometrical engineering enables the development of stretchable systems with a stable structure and limited stretchability. Recent studies have also explored the combination of both intrinsically stretchable materials and geometrical engineering of intrinsically stretchable material to attain improved electrical and mechanical stability. Besides material development, it is critical to develop a stretchable platform to enable the integration of devices. In this regard, fully stretchable systems and semi-stretchable systems are the two broad techniques which utilise either intrinsically stretchable devices or high-performance inorganic rigid devices connected with stretchable interconnects. Intrinsically stretchable devices with elastomeric material have the capability to handle large strain, but the performance and device mobility is a challenging issue. To achieve high device performance, inorganic materials are tailored to wavy or serpentine geometries which enable limited stretchability along with high device performance. The major barrier in the commercialisation of a stretchable system lies in the development of the large-scale manufacturing and integration techniques with assembly of multiple devices on a single platform. So far, large-area integration has been reported with stretchable interconnects that connect the rigid island of high-performance devices. The integration of an intrinsically stretchable inorganic device with stretchable interconnects has the potential to advance the stretchable system to the next level in terms of stretchability and performance.

References

[1] A. M. Zamarayeva, A. E. Ostfeld, M. Wang *et al.*, 'Flexible and Stretchable Power Sources for Wearable Electronics', *Sci. Adv.*, vol. 3, no. 6, p. e1602051, 2017.

[2] L. Manjakkal, L. Yin, A. Nathan, J. Wang and R. Dahiya, 'Energy Autonomous Sweat Based Wearable Systems', *Adv. Mater.*, 2021 (DOI: 10.1002/adma.202100899).

[3] J. Kim, H. J. Shim, J. Yang *et al.*, 'Ultrathin Quantum Dot Display Integrated with Wearable Electronics', *Adv. Mater.*, vol. 29, no. 38, p. 1700217, 2017.

[4] J. A. Rogers, 'Nanomesh On-Skin Electronics', *Nat. Nanotechnol.*, vol. 12, no. 9, pp. 839–40, 2017.

[5] P. Escobedo, M. Ntagios, D. Shakthivel, W. T. Navaraj and R. Dahiya, 'Energy Generating Electronic Skin with Intrinsic Touch Sensing', *IEEE Trans. Robot.*, vol. 37, no. 2, pp. 683–90, 2021.

[6] P. Escobedo, M. Bhattacharjee, F. Nikbakhtnasrabadi and R. Dahiya, 'Smart Bandage with Wireless Strain and Temperature Sensors and Battery-less NFC Tag', *IEEE Internet Things J.*, vol. 8, no.6, pp. 5093–100, 2021.

[7] W. Zhou, S. Yao, H. Wang *et al.*, 'Gas-Permeable, Ultrathin, Stretchable Epidermal Electronics with Porous Electrodes', *ACS Nano*, vol. 14, no. 5, pp. 5798–805, 2020.

[8] J.-H. Kim, S.-R. Kim, H.-J. Kil, Y.-C. Kim and J.-W. Park, 'Highly Conformable, Transparent Electrodes for Epidermal Electronics', *Nano Lett.*, vol. 18, no. 7, pp. 4531–40, 2018.

[9] Y. Kumaresan, O. Ozioko and R. Dahiya, 'Multifunctional Sensorized Electronic Skin to Detect and Distinguish Pressure and Temperature Stimuli', *IEEE Sens. J.*, 2021 (DOI: 10.1109/JSEN.2021.3055458).

[10] L. Manjakkal, W. Dang, N. Yogeswaran and R. Dahiya, 'Textile-Based Potentiometric Electrochemical pH Sensor for Wearable Applications', *Biosensors*, vol. 9, no. 1, pp. 0–12, 2019.

[11] R. Dahiya, D. Akinwande and J. S. Chang, 'Flexible Electronic Skin: From Humanoids to Humans', *Proc. IEEE*, vol. 107, no. 10, pp. 2011–15, 2019.

[12] W. Dang, V. Vinciguerra, L. Lorenzelli and R. Dahiya, 'Printable Stretchable Interconnects', *Flex. Print. Electron.*, vol. 2, no. 1, p. 013003, 2017.

[13] R. Mukherjee, P. Ganguli and R. Dahiya, 'Bioinspired Distributed Energy in Robotics and Enabling Technologies', *Adv. Intell. Syst.*, 2021 (DOI: 10.1002/aisy.202100036).

[14] W. Wu, 'Stretchable Electronics: Functional Materials, Fabrication Strategies and Applications', *Sci. Technol. Adv. Mater.*, vol. 20, no. 1, pp. 187–224, 2019.

[15] M. Amjadi, K.-U. Kyung, I. Park and M. Sitti, 'Stretchable, Skin-Mountable, and Wearable Strain Sensors and Their Potential Applications: A Review', *Adv. Funct. Mater.*, vol. 26, no. 11, pp. 1678–98, 2016.

[16] E. S. Hosseini, S. Dervin, P. Ganguly and R. Dahiya, 'Biodegradable Materials for Sustainable Health Monitoring Devices', *ACS Appl. Bio Mater.*, vol. 4, no. 1, pp. 163–94, 2021.

[17] D.-H. Kim and J. A. Rogers, 'Stretchable Electronics: Materials Strategies and Devices', *Adv. Mater.*, vol. 20, no. 24, pp. 4887–92, 2008.

[18] R. S. Dahiya, 'Epidermal Electronics – Flexible Electronics for Biomedical Applications', in *Handbook of Bioelectronics: Directly Interfacing Electronics and Biological Systems*, K. Iniewski and S. Carrara, eds. Cambridge: Cambridge University Press, 2015, pp. 245–55.

[19] C. García Núñez, L. Manjakkal and R. Dahiya, 'Energy Autonomous Electronic Skin', *npj Flex. Electron.*, vol. 3, no. 1, p. 1, 2019.

[20] M. Soni and R. Dahiya, 'Soft eSkin: Distributed Touch Sensing with Harmonized Energy and Computing', *Philos. T. R. Soc. A*, vol. 378, no. 2164, p. 20190156, 2020.

[21] R. Dahiya, 'E-Skin: From Humanoids to Humans [Point of View]', *Proc. IEEE*, vol. 107, no. 2, pp. 247–52, 2019.

[22] R. Dahiya, N. Yogeswaran, F. Liu *et al.*, 'Large-Area Soft e-Skin: The Challenges Beyond Sensor Designs', *Proc. IEEE*, vol. 107, no. 10, pp. 2016–33, 2019.

[23] L. Manjakkal, A. Pullanchiyodan, N. Yogeswaran, E. S. Hosseini and R. Dahiya, 'A Wearable Supercapacitor Based on Conductive PEDOT: PSS-Coated Cloth and a Sweat Electrolyte', *Adv. Mater.*, vol. 32, no. 24, p. 1907254, 2020.

[24] A. Pullanchiyodan, L. Manjakkal, S. Dervin, D. Shakthivel and R. Dahiya, 'Metal Coated Conductive Fabrics with Graphite Electrodes and Biocompatible Gel Electrolyte for Wearable Supercapacitors', *Adv. Mater. Technol.*, vol. 5, no. 5, p. 1901107, 2020.

[25] L. Manjakkal, W. T. Navaraj, C. García Núñez and R. Dahiya, 'Graphene–Graphite Polyurethane Composite Based High-Energy Density Flexible Supercapacitors', *Adv. Sci.*, vol. 6, no. 7, p. 1802251, 2019.

[26] L. Manjakkal, F. F. Franco, A. Pullanchiyodan, M. G. Jimenez and R. Dahiya, 'Natural Jute Fibre based Supercapacitors and Sensors for Eco-friendly Energy Autonomous Systems', *Adv. Sustain. Syst.*, vol. 5, no. 3, p. 2000286, 2021.

[27] A. Savov, S. Joshi, S. Shafqat *et al.*, 'A Platform for Mechano(-electrical) Characterization of Free-standing Micron- Sized Structures and Interconnects', *Micromachines*, vol. 9, no. 1, p. 39, 2018.

[28] D.-H. Kim, J. Song, W. M. Choi *et al.*, 'Materials and Noncoplanar Mesh Designs for Integrated Circuits with Linear Elastic Responses to Extreme Mechanical Deformations', *Proc. Natl. Acad. Sci.*, vol. 105, no. 48, p. 18675, 2008.

[29] M. A. Darabi, A. Khosrozadeh, Q. Wang and M. Xing, 'Gum Sensor: A Stretchable, Wearable, and Foldable Sensor Based on Carbon Nanotube/Chewing Gum Membrane', *ACS Appl. Mater. Interfaces*, vol. 7, no. 47, pp. 26195–205, 2015.

[30] Y. Zhou, C. Zhao, J. Wang *et al.*, 'Stretchable High-Permittivity Nanocomposites for Epidermal Alternating-Current Electroluminescent Displays', *ACS Mater. Lett.*, vol. 1, no. 5, pp. 511–18, 2019.

[31] T. Q. Trung, V. Q. Dang, H.-B. Lee *et al.*, 'An Omnidirectionally Stretchable Photodetector Based on Organic–Inorganic Heterojunctions', *ACS Appl. Mater. Interfaces*, vol. 9, no. 41, pp. 35958–67, 2017.

[32] H.-W. Jang, S. Kim and S.-M. Yoon, 'Impact of Polyimide Film Thickness for Improving the Mechanical Robustness of Stretchable InGaZnO Thin-Film Transistors Prepared on Wavy-Dimensional Elastomer Substrates', *ACS Appl. Mater. Interfaces*, vol. 11, no. 37, pp. 34076–83, 2019.

[33] C. W. Bae, P. T. Toi, B. Y. Kim *et al.*, 'Fully Stretchable Capillary Microfluidics-Integrated Nanoporous Gold Electrochemical Sensor for Wearable Continuous Glucose Monitoring', *ACS Appl. Mater. Interfaces*, vol. 11, no. 16, pp. 14567–75, 2019.

[34] N. Münzenrieder, G. Cantarella, C. Vogt *et al.*, 'Stretchable and Conformable Oxide Thin-Film Electronics', *Adv. Electron. Mater.*, vol. 1, no. 3, p. 1400038, 2015.

[35] B.-U. Hwang, A. Zabeeb, T. Q. Trung *et al.*, 'A Transparent Stretchable Sensor for Distinguishable Detection of Touch and Pressure by Capacitive and Piezoresistive Signal Transduction', *NPG Asia Mater.*, vol. 11, no. 1, p. 23, 2019.

[36] M. Jo, S. Bae, I. Oh *et al.*, '3D Printer-Based Encapsulated Origami Electronics for Extreme System Stretchability and High Areal Coverage', *ACS Nano*, vol. 13, no. 11, pp. 12500–10, 2019.

[37] A. J. Bandodkar, J.-M. You, N.-H. Kim *et al.*, 'Soft, Stretchable, High Power Density Electronic Skin-Based Biofuel Cells for Scavenging Energy from Human Sweat', *Energy Environ. Sci.*, vol. 10, no. 7, pp. 1581–9, 2017.

[38] A. M. V. Mohan, N. Kim, Y. Gu *et al.*, 'Merging of Thin- and Thick-Film Fabrication Technologies: Toward Soft Stretchable "Island–Bridge" Devices', *Adv. Mater. Technol.*, vol. 2, no. 4, p. 1600284, 2017.

[39] S. Yao, P. Ren, R. Song *et al.*, 'Nanomaterial-Enabled Flexible and Stretchable Sensing Systems: Processing, Integration, and Applications', *Adv. Mater.*, vol. 32, no. 15, p. 1902343, 2020.

[40] R. Tang and H. Fu, 'Mechanics of Buckled Kirigami Membranes for Stretchable Interconnects in Island–Bridge Structures', *J. Appl. Mech.*, vol. 87, no. 5, p. 051002, 2020.

[41] R. Dahiya, W. T. Navaraj, S. Khan and E. O. Polat, 'Developing Electronic Skin with the Sense of Touch', *Information Display*, vol. 31, no. 4, pp. 6–10, 2015.

[42] J. Zhou, G. Tian, G. Jin *et al.*, 'Buckled Conductive Polymer Ribbons in Elastomer Channels as Stretchable Fiber Conductor', *Adv. Funct. Mater.*, vol. 30, no. 5, p. 1907316, 2020.

[43] C. Jiang, Q. Li, S. Fan *et al.*, 'Hyaline and Stretchable Haptic Interfaces Based on Serpentine-Shaped Silver Nanofiber Networks', *Nano Energy*, vol. 73, p. 104782, 2020.

[44] P. Li, W. Zhang, J. Ma *et al.*, 'Solution-Grown Serpentine Silver Nanofiber Meshes for Stretchable Transparent Conductors', *Adv. Electron. Mater.*, vol. 4, no. 12, p. 1800346, 2018.

[45] Y. Zhang, M. Li, B. Qin *et al.*, 'Highly Transparent, Underwater Self-Healing, and Ionic Conductive Elastomer Based on Multivalent Ion–Dipole Interactions', *Chem. Mater.*, vol. 32, no. 15, pp. 6310–17, 2020.

[46] H. Stoyanov, M. Kollosche, S. Risse, R. Waché and G. Kofod, 'Soft Conductive Elastomer Materials for Stretchable Electronics and Voltage Controlled Artificial Muscles', *Adv. Mater.*, vol. 25, no. 4, pp. 578–83, 2013.

[47] W. Dang, V. Vinciguerra, L. Lorenzelli and R. Dahiya, 'Metal–Organic Dual Layer Structure for Stretchable Interconnects', *Procedia Eng.*, vol. 168, pp. 1559–62, 2016.

[48] H. Sun, Z. Han and N. Willenbacher, 'Ultrastretchable Conductive Elastomers with a Low Percolation Threshold for Printed Soft Electronics', *ACS Appl. Mater. Interfaces*, vol. 11, no. 41, pp. 38092–102, 2019.

[49] P. Lee, J. Lee, H. Lee *et al.*, 'Highly Stretchable and Highly Conductive Metal Electrode by Very Long Metal Nanowire Percolation Network', *Adv. Mater.*, vol. 24, no. 25, pp. 3326–32, 2012.

[50] W. Dang, V. Vinciguerra, L. Lorenzelli and R. Dahiya, 'Printable Stretchable Interconnects', *Flexible and Printed Electronics*, vol. 2, no. 1, p. 013003, 2017.

[51] C. W. Park, Y. G. Moon, H. Seong *et al.*, 'Photolithography-Based Patterning of Liquid Metal Interconnects for Monolithically Integrated Stretchable Circuits', *ACS Appl. Mater. Interfaces*, vol. 8, no. 24, pp. 15459–65, 2016.

[52] C. Lv, H. Yu and H. Jiang, 'Archimedean Spiral Design for Extremely Stretchable Interconnects', *Extreme Mech. Lett.*, vol. 1, pp. 29–34, 2014.

[53] R. Rahimi, M. Ochoa, W. Yu and B. Ziaie, 'A Sewing-Enabled Stitch-and-Transfer Method for Robust, Ultra-stretchable, Conductive Interconnects', *J. Micromech.Microeng.*, vol. 24, no. 9, p. 095018, 2014.

[54] R. Xu, M. Ochoa, W. Yu and B. Ziaie, 'Fabric-Based Stretchable Electronics with Mechanically Optimized Designs and Prestrained Composite Substrates', *Extreme Mech. Lett.*, vol. 1, pp. 120–6, 2014.

[55] Y. Zhao, W. Yang, Y. J. Tan *et al.*, 'Highly Conductive 3D Metal-Rubber Composites for Stretchable Electronic Applications', *APL Mater.*, vol. 7, no. 3, p. 031508, 2019.

[56] Z. Xue, H. Song, J. A. Rogers, Y. Zhang and Y. Huang, 'Mechanically-Guided Structural Designs in Stretchable Inorganic Electronics', *Adv. Mater.*, vol. 32, no. 15, p. 1902254, 2019.

[57] T. Q. Trung and N.-E. Lee, 'Recent Progress on Stretchable Electronic Devices with Intrinsically Stretchable Components', *Adv. Mater.*, vol. 29, no. 3, p. 1603167, 2017.

[58] J.-H. Lee, K. Y. Lee, M. K. Gupta *et al.*, 'Highly Stretchable Piezoelectric-Pyroelectric Hybrid Nanogenerator', *Adv. Mater.*, vol. 26, no. 5, PP. 765–9, 2014.

[59] H. Wu, S. Kustra, E. M. Gates and C. J. Bettinger, 'Topographic Substrates as Strain Relief Features in Stretchable Organic Thin Film Transistors', *Org. Electron.*, vol. 14, no. 6, PP. 1636–42, 2013.

[60] D.-H. Kim, J.-H. Ahn, W. M. Choi *et al.*, 'Stretchable and Foldable Silicon Integrated Circuits', *Science*, vol. 320, no. 5875, p. 507, 2008.

[61] Y. Sun, V. Kumar, I. Adesida and J. A. Rogers, 'Buckled and Wavy Ribbons of GaAs for High- Performance Electronics on Elastomeric Substrates', *Adv. Mater.*, vol. 18, no. 21, pp. 2857–62, 2006.

[62] Y. Wang, Z. Li and J. Xiao, 'Stretchable Thin Film Materials: Fabrication, Application, and Mechanics', *J. Electron. Packag.*, vol. 138, no. 2, p. 020801, 2016.

[63] K. Park, D.-K. Lee, B.-S. Kim *et al.*, 'Stretchable, Transparent Zinc Oxide Thin Film Transistors', *Adv. Funct. Mater.*, vol. 20, no. 20, pp. 3577–82, 2010.

[64] M. Kaltenbrunner, M. S. White, E. D. Głowacki *et al.*, 'Ultrathin and Lightweight Organic Solar Cells with High Flexibility', *Nat. Commun.*, vol. 3, no. 1, p. 770, 2012.

[65] H. Zhou, W. Qin, Q. Yu *et al.*, 'Controlled Buckling and Postbuckling Behaviors of Thin Film Devices Suspended on an Elastomeric Substrate with Trapezoidal Surface Relief Structures', *Int. J. Solids and Struct.*, vol. 160, pp. 96–102, 2019.

[66] A. J. Baca, J.-H. Ahn, Y. Sun *et al.*, 'Semiconductor Wires and Ribbons for High-Performance Flexible Electronics', *Angew. Chem.*, vol. 47, no. 30, pp. 5524–42, 2008.

[67] Y. Duan, Y. Huang, Z. Yin, N. Bu and W. Dong, 'Non-wrinkled, Highly Stretchable Piezoelectric Devices by Electrohydrodynamic Direct-Writing', *Nanoscale*, vol. 6, no. 6, pp. 3289–95, 2014.

[68] C. Zhao, X. Jia, K. Shu *et al.* et al., 'Stretchability Enhancement of Buckled Polypyrrole Electrode for Stretchable Supercapacitors via Engineering Substrate Surface Roughness', *Electrochimic.Acta*, vol. 343, p. 136099, 2020.

[69] B. Wang, S. Bao, S. Vinnikova, P. Ghanta and S. Wang, 'Buckling Analysis in Stretchable Electronics', *npj Flex. Electron.*, vol. 1, no. 1, p. 5, 2017.

[70] H. Jiang, D.-Y. Khang, J. Song *et al.*, 'Finite Deformation Mechanics in Buckled Thin Films on Compliant Supports', *Proc. Natl. Acad. Sci.*, vol. 104, no. 40, p. 15607, 2007.

[71] Y. Zhang, Y. Huang and J. A. Rogers, 'Mechanics of Stretchable Batteries and Supercapacitors', *Curr. Opin. Solid State Mater. Sci.*, vol. 19, no. 3, pp. 190–9, 2015.

[72] Z. Y. Huang, W. Hong and Z. Suo, 'Nonlinear Analyses of Wrinkles in a Film Bonded to a Compliant Substrate', *J. Mech. Phys. Solids*, vol. 53, no. 9, PP. 2101–18, 2005.

[73] D.-Y. Khang, J. A. Rogers and H. H. Lee, 'Mechanical Buckling: Mechanics, Metrology, and Stretchable Electronics', *Adv. Funct. Mater.*, vol. 19, no. 10, PP. 1526–36, 2009.

[74] H. Jiang, D.-Y. Khang, H. Fei *et al.*, 'Finite Width Effect of Thin-Films Buckling on Compliant Substrate: Experimental and Theoretical Studies', *J. Mech. Phys. Solids*, vol. 56, no. 8, PP. 2585–98, 2008.

[75] S. Wang, J. Xu, W. Wang *et al.*, 'Skin Electronics from Scalable Fabrication of an Intrinsically Stretchable Transistor Array', *Nature*, vol. 555, no. 7694, PP. 83–8, 2018.

[76] D. C. Kim, H. J. Shim, W. Lee, J. H. Koo and D.-H. Kim, 'Material-Based Approaches for the Fabrication of Stretchable Electronics', *Adv. Mater.*, vol. 32, no. 15, p. 1902743, 2020.

[77] H.-C. Wu, S. J. Benight, A. Chortos *et al.*, 'A Rapid and Facile Soft Contact Lamination Method: Evaluation of Polymer Semiconductors for Stretchable Transistors', *Chem. Mater.*, vol. 26, no. 15, PP. 4544–51, 2014.

[78] H.-C. Wu, C.-C. Hung, C.-W. Hong *et al.*, 'Isoindigo-Based Semiconducting Polymers Using Carbosilane Side Chains for High Performance Stretchable Field-Effect Transistors', *Macromolecules*, vol. 49, no. 22, PP. 8540–8, 2016.

[79] C.-H. Li, C. Wang, C. Keplinger *et al.*, 'A Highly Stretchable Autonomous Self-healing Elastomer', *Nat. Chem*, vol. 8, no. 6, PP. 618–24, 2016.

[80] M. Ashizawa, Y. Zheng, H. Tran and Z. Bao, 'Intrinsically Stretchable Conjugated Polymer Semiconductors in Field Effect Transistors', *Prog. Polym. Sci.*, vol. 100, p. 101181, 2020.

[81] K. Sim, Z. Rao, H.-J. Kim *et al.*, 'Fully Rubbery Integrated Electronics from High Effective Mobility Intrinsically Stretchable Semiconductors', *Sci. Adv.*, vol. 5, no. 2, p. eaav5749, 2019.

[82] S. Ding, Z. Jiang, F. Chen *et al.*, 'Intrinsically Stretchable, Transient Conductors from a Composite Material of Ag Flakes and Gelatin Hydrogel', *ACS Appl. Mater. Interfaces*, vol. 12, no. 24, PP. 27572–7, 2020.

[83] J. Liang, L. Li, K. Tong *et al.*, 'Silver Nanowire Percolation Network Soldered with Graphene Oxide at Room Temperature and Its Application for Fully Stretchable Polymer Light-Emitting Diodes', *ACS Nano*, vol. 8, no. 2, PP. 1590–600, 2014.

[84] A. Chortos, C. Zhu, J. Y. Oh *et al.*, 'Investigating Limiting Factors in Stretchable All-Carbon Transistors for Reliable Stretchable Electronics', *ACS Nano*, vol. 11, no. 8, PP. 7925–37, 2017.

[85] K.-I. Jang, K. Li, H. U. Chung *et al.*, 'Self-assembled Three Dimensional Network Designs for Soft Electronics', *Nat. Commun.*, vol. 8, no. 1, p. 15894, 2017.

[86] M. U. Rehman and J. P. Rojas, 'Optimization of Compound Serpentine–Spiral Structure for Ultra-stretchable Electronics', *Extreme Mech. Lett.*, vol. 15, PP. 44–50, 2017.

[87] J. Woo, H. Lee, C. Yi *et al.*, 'Ultrastretchable Helical Conductive Fibers Using Percolated Ag Nanoparticle Networks Encapsulated by Elastic Polymers with High Durability in Omnidirectional Deformations for Wearable Electronics', *Adv. Funct. Mater.*, vol. 30, no. 29, p. 1910026, 2020.

[88] C. Deng, L. Pan, C. Li, X. Fu, R. Cui and H. Nasir, 'Helical Gold Nanotube Film as Stretchable Micro/Nanoscale Strain Sensor', *J. Mater. Sci.*, vol. 53, no. 3, PP. 2181–92, 2018.

[89] Z. Xie, R. Avila, Y. Huang and J. A. Rogers, 'Flexible and Stretchable Antennas for Biointegrated Electronics', *Adv. Mater.*, vol. 32, no. 15, p. 1902767, 2019.

[90] E. Bonderover and S. Wagner, 'A Woven Inverter Circuit for e-textile Applications', *IEEE Electron. Device Lett.*, vol. 25, no. 5, PP. 295–7, 2004.

[91] M. Hamedi, R. Forchheimer and O. Inganäs, 'Towards Woven Logic from Organic Electronic Fibres', *Nat. Mater.*, vol. 6, no. 5, PP. 357–62, 2007.

[92] E. Kang, W. Min, H. Choo and J.-E. Park, 'Design of a Very High Frequency Stretchable Inverted Conical Helical Antenna for Maritime Search and Rescue Applications', *Microw. Opt. Technol. Lett.*, vol. 62, no. 1, PP. 284–8, 2020.

[93] T. N. Do and Y. Visell, 'Stretchable, Twisted Conductive Microtubules for Wearable Computing, Robotics, Electronics, and Healthcare', *Sci. Rep.*, vol. 7, no. 1, p. 1753, 2017.

[94] Y. Tai and G. Lubineau, 'Double-Twisted Conductive Smart Threads Comprising a Homogeneously and a Gradient-Coated Thread for Multidimensional Flexible Pressure-Sensing Devices', *Adv. Funct. Mater.*, vol. 26, no. 23, pp. 4078–84, 2016.

[95] Y. H. Jung, J. Lee, Y. Qiu *et al.*, 'Stretchable Twisted-Pair Transmission Lines for Microwave Frequency Wearable Electronics', *Adv. Funct. Mater.*, vol. 26, no. 26, pp. 4635–42, 2016.

[96] M. Y. Cheng, C. M. Tsao, Y. Z. Lai and Y. J. Yang, 'The Development of a Highly Twistable Tactile Sensing Array with Stretchable Helical Electrodes', *Sensor. Actuat. A-Phys.*, vol. 166, no. 2, pp. 226–33, 2011.

[97] W. Dang, L. Manjakkal, W. T. Navaraj, L. Lorenzelli, V. Vinciguerra and R. Dahiya, 'Stretchable Wireless System for Sweat pH Monitoring', *Biosens. Bioelectron.*, vol. 107, pp. 192–202, 2018.

[98] H. Hocheng and C.-M. Chen, 'Design, Fabrication and Failure Analysis of Stretchable Electrical Routings', *Sensors*, vol. 14, no. 7, pp. 11855–77, 2014.

[99] M. Cheng, C. Tsao and Y. Yang, 'An Anthropomorphic Robotic Skin Using Highly Twistable Tactile Sensing Array', in *2010 5th IEEE Conference on Industrial Electronics and Applications*, pp. 650–5, 2010 (DOI: 10.1109/ICIEA.2010.5517008).

[100] B. Huyghe, H. Rogier, J. Vanfleteren and F. Axisa, 'Design and Manufacturing of Stretchable High-Frequency Interconnects', *IEEE Trans. Adv. Packag.*, vol. 31, no. 4, pp. 802–8, 2008.

[101] D. Brosteaux, F. Axisa, M. Gonzalez and J. Vanfleteren, 'Design and Fabrication of Elastic Interconnections for Stretchable Electronic Circuits', *IEEE Electron. Device Lett.*, vol. 28, no. 7, pp. 552–4, 2007.

[102] R. Verplancke, F. Bossuyt, D. Cuypers and J. Vanfleteren, 'Thin-Film Stretchable Electronics Technology Based on Meandering Interconnections: Fabrication and Mechanical Performance', *J. Micromech.Microeng.*, vol. 22, no. 1, p. 015002, 2011.

[103] D.-H. Kim, Z. Liu, Y.-S. Kim *et al.*, 'Optimized Structural Designs for Stretchable Silicon Integrated Circuits', *Small*, vol. 5, no. 24, pp. 2841–7, 2009.

[104] Y. Zhang, H. Fu, Y. Su *et al.*, 'Mechanics of Ultra-stretchable Self-similar Serpentine Interconnects', *Acta Mater.*, vol. 61, no. 20, pp. 7816–27, 2013.

[105] R.-H. Kim, D.-H. Kim, J. Xiao *et al.*, 'Waterproof AlInGaP Optoelectronics on Stretchable Substrates with Applications in Biomedicine and Robotics', *Nat. Mater.*, vol. 9, no. 11, pp. 929–37, 2010.

[106] R. C. Webb, A. P. Bonifas, A. Behnaz *et al.*, 'Ultrathin Conformal Devices for Precise and Continuous Thermal Characterization of Human Skin', *Nat. Mater.*, vol. 12, no. 10, pp. 938–44, 2013.

[107] Z. Fan, Y. Zhang, Q. Ma *et al.*, 'A Finite Deformation Model of Planar Serpentine Interconnects for Stretchable Electronics', *Int. J. Solids and Struct.*, vol. 91, pp. 46–54, 2016.

[108] W. Dang, 'Stretchable Interconnects for Smart Integration of Sensors in Wearable and Robotic Applications', doctorate, School of Engineering, University of Glasgow, glathesis:2018-40994, 2018.

[109] R. Xu, Y. Zhang and K. Komvopoulos, 'Mechanical Designs Employing Buckling Physics for Reversible and Omnidirectional Stretchability in Microsupercapacitor Arrays', *Mater. Res. Lett.*, vol. 7, no. 3, pp. 110–16, 2019.

[110] M. Isobe and K. Okumura, 'Initial Rigid Response and Softening Transition of Highly Stretchable Kirigami Sheet Materials', *Sci. Rep.*, vol. 6, no. 1, p. 24758, 2016.

[111] C. Zhang, A. Khan, J. Cai *et al.*, 'Stretchable Transparent Electrodes with Solution-Processed Regular Metal Mesh for an Electroluminescent Light-Emitting Film', *ACS Appl. Mater. Interfaces*, vol. 10, no. 24, pp. 21009–17, 2018.

[112] H. C. Lee, E. Y. Hsieh, K. Yong and S. Nam, 'Multiaxially-Stretchable Kirigami-Patterned Mesh Design for Graphene Sensor Devices', *Nano Res.*, vol. 13, no. 5, pp. 1406–12, 2020.

[113] Y. Morikawa, S. Ayub, O. Paul, T. Kawano and P. Ruther, 'Highly Stretchable Kirigami Structure with Integrated Led Chips and

Electrodes for Optogenetic Experiments on Perfused Hearts', in *2019 20th International Conference on Solid-State Sensors, Actuators and Microsystems & Eurosensors XXXIII (TRANSDUCERS & EUROSENSORS XXXIII)*, pp. 2484–7, 2019 (DOI: 10.1109/ TRANSDUCERS.2019.8808221).

[114] R. Xu, A. Hung, A. Zverev *et al.*, 'A Kirigami-Inspired, Extremely Stretchable, High Areal-Coverage Micro-supercapacitor Patch', in *2018 IEEE Micro. Electro. Mechanical Systems (MEMS)*, pp. 661–4, 2018 (DOI: 10.1109/MEMSYS.2018.8346641).

[115] C. Wu, X. Wang, L. Lin, H. Guo and Z. L. Wang, 'Paper-Based Triboelectric Nanogenerators Made of Stretchable Interlocking Kirigami Patterns', *ACS Nano*, vol. 10, no. 4, pp. 4652–9, 2016.

[116] N.-S. Jang, K.-H. Kim, S.-H. Ha, S.-H. Jung, H. M. Lee and J.-M. Kim, 'Simple Approach to High- Performance Stretchable Heaters Based on Kirigami Patterning of Conductive Paper for Wearable Thermotherapy Applications', *ACS Appl. Mater. Interfaces*, vol. 9, no. 23, pp. 19612–21, 2017.

[117] Z. Wang, L. Zhang, S. Duan, H. Jiang, J. Shen and C. Li, 'Kirigami-Patterned Highly Stretchable Conductors from Flexible Carbon Nanotube-Embedded Polymer Films', *J. Mater. Chem. C*, vol. 5, no. 34, pp. 8714–22, 2017.

[118] R. Xu, A. Zverev, A. Hung *et al.*, 'Kirigami-Inspired, Highly Stretchable Micro-supercapacitor Patches Fabricated by Laser Conversion and Cutting', *Microsyst. Nanoeng.*, vol. 4, no. 1, p. 36, 2018.

[119] Z. Zhang, Y. Yu, Y. Tang *et al.*, 'Kirigami-Inspired Stretchable Conjugated Electronics', *Adv. Electron. Mater.*, vol. 6, no. 1, p. 1900929, 2020.

[120] J. Qi, H. Xiong, C. Hou, Q. Zhang, Y. Li and H. Wang, 'A Kirigami-Inspired Island-Chain Design for Wearable Moistureproof Perovskite Solar Cells with High Stretchability and Performance Stability', *Nanoscale*, vol. 12, no. 6, pp. 3646–56, 2020.

[121] Y. Bao, G. Hong, Y. Chen *et al.*, 'Customized Kirigami Electrodes for Flexible and Deformable Lithium-Ion Batteries', *ACS Appl. Mater. Interfaces*, vol. 12, no. 1, pp. 780–8, 2020.

[122] M. Isobe and K. Okumura, 'Discontinuity in the In-plane to Out-of-plane Transition of Kirigami', *J. Phys. Soc. Japan*, vol. 88, no. 2, p. 025001, 2019.

[123] B.-Y. Kim, H.-B. Lee and N.-E. Lee, 'A Durable, Stretchable, and Disposable Electrochemical Biosensor on Three-Dimensional Micro-patterned Stretchable Substrate', *Sensor. Actuat. B- Chem.*, vol. 283, pp. 312–20, 2019.

[124] H.-B. Lee, C.-W. Bae, L. T. Duy *et al.*, 'Mogul-Patterned Elastomeric Substrate for Stretchable Electronics', *Adv. Mater.*, vol. 28, no. 16, pp. 3069–77, 2016.

[125] Y.-H. Lee and Y.-J. Kim, 'Mechanical Characteristics of Stretchable Electronics Based on a Mogul-Patterned Structure', *Funct. Mater. Lett.*, vol. 9, no. 6, p. 1642012, 2016.

[126] C. Wu, J. Jiu, T. Araki *et al.*, 'Biaxially Stretchable Silver Nanowire Conductive Film Embedded in a Taro Leaf- Templated PDMS Surface', *Nanotechnology*, vol. 28, no. 1, p. 01LT01, 2016.

[127] T. Takahashi, K. Takei, A. G. Gillies, R. S. Fearing and A. Javey, 'Carbon Nanotube Active-Matrix Backplanes for Conformal Electronics and Sensors', *Nano Lett.*, vol. 11, no. 12, pp. 5408–13, 2011.

[128] J.-E. Lim, S. Yoon, B.-U. Hwang, N.-E. Lee and H.-K. Kim, 'Self-connected Ag Nanoporous Sponge Embedded in Sputtered Polytetrafluoroethylene for Highly Stretchable and Semi-transparent Electrodes', *Adv. Mater. Interfaces*, vol. 6, no. 8, p. 1801936, 2019.

[129] X. Gui, A. Cao, J. Wei *et al.*, 'Soft, Highly Conductive Nanotube Sponges and Composites with Controlled Compressibility', *ACS Nano*, vol. 4, no. 4, pp. 2320–6, 2010.

[130] Z. Chen, W. Ren, L. Gao, B. Liu, S. Pei and H.-M. Cheng, 'Three-Dimensional Flexible and Conductive Interconnected Graphene Networks Grown by Chemical Vapour Deposition', *Nat. Mater.*, vol. 10, no. 6, pp. 424–8, 2011.

[131] Y. Yu, J. Zeng, C. Chen *et al.*, 'Three-Dimensional Compressible and Stretchable Conductive Composites', *Adv. Mater.*, vol. 26, no. 5, pp. 810–15, 2014.

[132] C. F. Guo, T. Sun, Q. Liu, Z. Suo and Z. Ren, 'Highly Stretchable and Transparent Nanomesh Electrodes Made by Grain Boundary Lithography', *Nat. Commun.*, vol. 5, no. 1, p. 3121, 2014.

[133] D.-H. Kim, N. Lu, R. Ma *et al.*, 'Epidermal Electronics', *Science*, vol. 333, no. 6044, p. 838, 2011.

[134] M. Bhattacharjee, M. Soni, P. Escobedo and R. Dahiya, 'PEDOT:PSS Microchannel-Based Highly Sensitive Stretchable Strain Sensor', *Adv. Electron. Mater.*, vol. 6, no. 8, p. 2000445, 2020.

[135] S. G. R. Bade, X. Shan, P. T. Hoang *et al.*, 'Stretchable Light-Emitting Diodes with Organometal-Halide-Perovskite– Polymer Composite Emitters', *Adv. Mater.*, vol. 29, no. 23, p. 1607053, 2017.

[136] T. Sekitani, H. Nakajima, H. Maeda *et al.*, 'Stretchable Active-Matrix Organic Light-Emitting Diode Display Using Printable Elastic Conductors', *Nat. Mater.*, vol. 8, no. 6, pp. 494–9, 2009.

[137] M.-S. Lee, K. Lee, S.-Y. Kim *et al.*, 'High-Performance, Transparent and Stretchable Electrodes Using Graphene–Metal Nanowire Hybrid Structures', *Nano Lett.*, vol. 13, no. 6, pp. 2814–21, 2013.

[138] G. S. Jeong, D.-H. Baek, H. C. Jung *et al.*, 'Solderable and Electroplatable Flexible Electronic Circuit on a Porous Stretchable Elastomer', *Nat. Commun.*, vol. 3, no. 1, p. 977, 2012.

[139] Y. Fouillet, C. Parent, G. Gropplero *et al.*, 'Stretchable Material for Microfluidic Applications', *MDPI Proc.*, vol. 1, no. 4, p. 501, 2017.

[140] S. Gupta, F. Carrillo, C. Li, L. Pruitt and C. Puttlitz, 'Adhesive Forces Significantly Affect Elastic Modulus Determination of Soft Polymeric Materials in Nanoindentation', *Mater. Lett.*, vol. 61, no. 2, pp. 448–51, 2007.

[141] S. H. Jeong, S. Zhang, K. Hjort, J. Hilborn and Z. Wu, 'PDMS-Based Elastomer Tuned Soft, Stretchable, and Sticky for Epidermal Electronics', *Adv. Mater.*, vol. 28, no. 28, pp. 5830–6, 2016.

[142] F. Schneider, T. Fellner, J. Wilde and U. Wallrabe, 'Mechanical Properties of Silicones for MEMS', *J. Micromech. Microeng.*, vol. 18, no. 6, p. 065008, 2008.

[143] K. Ma, J. Rivera, G. J. Hirasaki and S. L. Biswal, 'Wettability Control and Patterning of PDMS Using UV–Ozone and Water Immersion', *J. Colloid Interface Sci.*, vol. 363, no. 1, pp. 371–8, 2011.

[144] A. Oláh, H. Hillborg and G. J. Vancso, 'Hydrophobic Recovery of UV/Ozone Treated Poly(dimethylsiloxane): Adhesion Studies by Contact Mechanics and Mechanism of Surface Modification', *Appl. Surf. Sci.*, vol. 239, no. 3, pp. 410–23, 2005.

[145] D.-P. Wang, J.-C. Lai, H.-Y. Lai *et al.*, 'Distinct Mechanical and Self-healing Properties in Two Polydimethylsiloxane Coordination Polymers with Fine-Tuned Bond Strength', *Inorg. Chem.*, vol. 57, no. 6, pp. 3232–42, 2018.

[146] D. Döhler, J. Kang, C. B. Cooper *et al.*, 'Tuning the Self-healing Response of Poly(dimethylsiloxane)-Based Elastomers', *ACS Appl. Polym. Mater.*, vol. 2, no. 9, pp. 4127–39, 2020.

[147] C. H. Lee, Y. Ma, K.-I. Jang *et al.*, 'Soft Core/Shell Packages for Stretchable Electronics', *Adv. Funct. Mater.*, vol. 25, no. 24, pp. 3698–704, 2015.

[148] K.-I. Jang, S. Y. Han, S. Xu *et al.*, 'Rugged and Breathable Forms of Stretchable Electronics with Adherent Composite Substrates for Transcutaneous Monitoring', *Nat. Commun.*, vol. 5, no. 1, p. 4779, 2014.

[149] W.-H. Yeo, Y.-S. Kim, J. Lee *et al.*, 'Multifunctional Epidermal Electronics Printed Directly onto the Skin', *Adv. Mater.*, vol. 25, no. 20, pp. 2773–8, 2013.

[150] R. Zhang, K. Huang, M. Zhu *et al.*, 'Corrosion Resistance of Stretchable Electrospun SEBS/PANi Micro-Nano Fiber Membrane', *Eur. Polym. J.*, vol. 123, p. 109394, 2020.

[151] S. Choi, J. Park, W. Hyun *et al.*, 'Stretchable Heater Using Ligand-Exchanged Silver Nanowire Nanocomposite for Wearable Articular Thermotherapy', *ACS Nano*, vol. 9, no. 6, pp. 6626–33, 2015.

[152] S. Song and Y. Zhang, 'Carbon Nanotube/Reduced Graphene Oxide Hybrid for Simultaneously Enhancing the Thermal Conductivity and Mechanical Properties of Styrene-butadiene Rubber', *Carbon*, vol. 123, pp. 158–67, 2017.

[153] F. Bossuyt, J. Guenther, T. Löher, M. Seckel, T. Sterken and J. de Vries, 'Cyclic Endurance Reliability of Stretchable Electronic Substrates', *Microelectron. Reliab.*, vol. 51, no. 3, pp. 628–35, 2011.

[154] H. Li, J.-T. Sun, C. Wang *et al.*, 'High Modulus, Strength and Toughness Polyurethane Elastomer Based on Unmodified Lignin', *ACS Sustain. Chem. Eng.*, vol. 5, no. 9, pp. 7942–9, 2017.

[155] S. Uman, A. Dhand and J. A. Burdick, 'Recent Advances in Shear-Thinning and Self-healing Hydrogels for Biomedical Applications', *J. Appl. Polym. Sci.*, vol. 137, no. 25, p. 48668, 2020.

[156] B. Xu, H. Jiang, H. Li, G. Zhang and Q. Zhang, 'High Strength Nanocomposite Hydrogel Bilayer with Bidirectional Bending and Shape Switching Behaviors for Soft Actuators', *RSC Adv.*, vol. 5, no. 17, pp. 13167–70, 2015.

[157] D. Son, J. Kang, O. Vardoulis *et al.*, 'An Integrated Self-healable Electronic Skin System Fabricated via Dynamic Reconstruction of a Nanostructured Conducting Network', *Nat. Nanotechnol.*, vol. 13, no. 11, pp. 1057–65, 2018.

[158] L. Cao, J. Fan, J. Huang and Y. Chen, 'A Robust and Stretchable Cross-linked Rubber Network with Recyclable and Self-healable Capabilities Based on Dynamic Covalent Bonds', *J. Mater. Chem. A*, vol. 7, no. 9, pp. 4922–33, 2019.

[159] M. P. Wolf, G. B. Salieb-Beugelaar and P. Hunziker, 'PDMS with Designer Functionalities – Properties, Modifications Strategies, and Applications', *Prog. Polym. Sci.*, vol. 83, pp. 97–134, 2018.

[160] J. Vaicekauskaite, P. Mazurek, S. Vudayagiri and A. L. Skov, 'Mapping the Mechanical and Electrical Properties of Commercial Silicone Elastomer Formulations for Stretchable Transducers', *J. Mater. Chem. C*, vol. 8, no. 4, pp. 1273–9, 2020.

[161] S. Park, K. Mondal, R. M. Treadway *et al.*, 'Silicones for Stretchable and Durable Soft Devices: Beyond Sylgard-184', *ACS Appl. Mater. Interfaces.*, vol. 10, no. 13, pp. 11261–8, 2018.

[162] *PUR Rubber*. https://designerdata.nl/materials/plastics/rubbers/polyur ethane-rubber.

[163] J. C. K. de Verney, M. F. S. Lima and D. M. Lenz, 'Properties of SBS and Sisal Fiber Composites: Ecological Material for Shoe Manufacturing', *Mater. Res.*, vol. 11, pp. 447–51, 2008.

[164] P. Gupta, M. Bera and P. K. Maji, 'Nanotailoring of Sepiolite Clay with Poly[styrene-b-(ethylene-co-butylene)-b-styrene]: Structure–Property Correlation', *Polym. Adv. Technol.*, vol. 28, no. 11, pp. 1428–37, 2017.

[165] N. Bowden, S. Brittain, A. G. Evans, J. W. Hutchinson and G. M. Whitesides, 'Spontaneous Formation of Ordered Structures in Thin Films of Metals Supported on an Elastomeric Polymer', *Nature*, vol. 393, no. 6681, pp. 146–9, 1998.

[166] S. P. Lacour, S. Wagner, Z. Huang and Z. Suo, 'Stretchable Gold Conductors on Elastomeric Substrates', *Appl. Phys. Lett.*, vol. 82, no. 15, pp. 2404–6, 2003.

[167] I. M. Graz, D. P. J. Cotton and S. P. Lacour, 'Extended Cyclic Uniaxial Loading of Stretchable Gold Thin-Films on Elastomeric Substrates', *Appl. Phys. Lett.*, vol. 94, no. 7, p. 071902, 2009.

[168] D. S. Gray, J. Tien and C. S. Chen, 'High-Conductivity Elastomeric Electronics', *Adv. Mater.*, vol. 16, no. 5, pp. 393–7, 2004.

[169] S. Khan, L. Lorenzelli and R. S. Dahiya, 'Technologies for Printing Sensors and Electronics over Large Flexible Substrates: A Review', *IEEE Sens. J.*, vol. 15, no. 6, pp. 3164–85, 2015.

[170] K.-Y. Chun, Y. Oh, J. Rho *et al.*, 'Highly Conductive, Printable and Stretchable Composite Films of Carbon Nanotubes and Silver', *Nat. Nanotechnol.*, vol. 5, no. 12, pp. 853–7, 2010.

[171] D. J. Finn, M. Lotya and J. N. Coleman, 'Inkjet Printing of Silver Nanowire Networks', *ACS Appl. Mater. Interfaces*, vol. 7, no. 17, pp. 9254–61, 2015.

[172] F. Michelis, L. Bodelot, Y. Bonnassieux and B. Lebental, 'Highly Reproducible, Hysteresis-Free, Flexible Strain Sensors by Inkjet Printing of Carbon Nanotubes', *Carbon*, vol. 95, pp. 1020–6, 2015.

[173] V. Scardaci, R. Coull, P. E. Lyons, D. Rickard and J. N. Coleman, 'Spray Deposition of Highly Transparent, Low-Resistance Networks of Silver Nanowires over Large Areas', *Small*, vol. 7, no. 18, pp. 2621–8, 2011.

[174] S. Khan, L. Lorenzelli and R. Dahiya, 'Flexible MISFET Devices from Transfer Printed Si Microwires and Spray Coating', *IEEE J. Electron. Devices Soc.*, vol. 4, no. 4, pp. 189–96, 2016.

[175] J. Liang, K. Tong and Q. Pei, 'A Water-Based Silver-Nanowire Screen-Print Ink for the Fabrication of Stretchable Conductors and Wearable Thin-Film Transistors', *Adv. Mater.*, vol. 28, no. 28, pp. 5986–96, 2016.

[176] S. Khan, W. Dang, L. Lorenzelli and R. Dahiya, 'Flexible Pressure Sensors Based on Screen- Printed P(VDF-TrFE) and P(VDF-TrFE)/MWCNTs', *IEEE Trans. Semicond. Manuf.*, vol. 28, no. 4, pp. 486–93, 2015.

[177] L. Yang, T. Zhang, H. Zhou, S. C. Price, B. J. Wiley and W. You, 'Solution-Processed Flexible Polymer Solar Cells with Silver Nanowire Electrodes', *ACS Appl. Mater. Interfaces*, vol. 3, no. 10, pp. 4075–84, 2011.

[178] T. Akter and W. S. Kim, 'Reversibly Stretchable Transparent Conductive Coatings of Spray- Deposited Silver Nanowires', *ACS Appl. Mater. Interfaces*, vol. 4, no. 4, pp. 1855–9, 2012.

[179] Z. Yu, Q. Zhang, L. Li *et al.*, 'Highly Flexible Silver Nanowire Electrodes for Shape-Memory Polymer Light- Emitting Diodes', *Adv. Mater.*, vol. 23, no. 5, pp. 664–8, 2011.

[180] S. De, T. M. Higgins, P. E. Lyons *et al.*, 'Silver Nanowire Networks as Flexible, Transparent, Conducting Films: Extremely High DC to Optical Conductivity Ratios', *ACS Nano*, vol. 3, no. 7, pp. 1767–74, 2009.

[181] P. Lee, J. Ham, J. Lee *et al.*, 'Highly Stretchable or Transparent Conductor Fabrication by a Hierarchical Multiscale Hybrid Nanocomposite', *Adv. Funct. Mater.*, vol. 24, no. 36, pp. 5671–8, 2014.

[182] J. Kim, M.-S. Lee, S. Jeon *et al.*, 'Highly Transparent and Stretchable Field-Effect Transistor Sensors Using Graphene–Nanowire Hybrid Nanostructures', *Adv. Mater.*, vol. 27, no. 21, pp. 3292–7, 2015.

[183] J.-Y. Lee, S. T. Connor, Y. Cui and P. Peumans, 'Solution-Processed Metal Nanowire Mesh Transparent Electrodes', *Nano Lett.*, vol. 8, no. 2, pp. 689–92, 2008.

[184] A. R. Madaria, A. Kumar, F. N. Ishikawa and C. Zhou, 'Uniform, Highly Conductive, and Patterned Transparent Films of a Percolating Silver Nanowire Network on Rigid and Flexible Substrates Using a Dry Transfer Technique', *Nano Res.*, vol. 3, no. 8, pp. 564–73, 2010.

[185] Y. Ko, S. K. Song, N. H. Kim and S. T. Chang, 'Highly Transparent and Stretchable Conductors Based on a Directional Arrangement of Silver Nanowires by a Microliter-Scale Solution Process', *Langmuir*, vol. 32, no. 1, pp. 366–73, 2016.

[186] A. R. Rathmell, M. Nguyen, M. Chi and B. J. Wiley, 'Synthesis of Oxidation-Resistant Cupronickel Nanowires for Transparent Conducting Nanowire Networks', *Nano Lett.*, vol. 12, no. 6, pp. 3193–9, 2012.

[187] M. Amjadi, A. Pichitpajongkit, S. Lee, S. Ryu and I. Park, 'Highly Stretchable and Sensitive Strain Sensor Based on Silver Nanowire–Elastomer Nanocomposite', *ACS Nano*, vol. 8, no. 5, pp. 5154–63, 2014.

[188] S. Kim, M. Amjadi, T.-I. Lee *et al.*, 'Wearable, Ultrawide-Range, and Bending-Insensitive Pressure Sensor Based on Carbon Nanotube Network-Coated Porous Elastomer Sponges for Human Interface and Healthcare Devices', *ACS Appl. Mater. Interfaces*, vol. 11, no. 26, pp. 23639–48, 2019.

[189] L. Ding, S. Xuan, L. Pei *et al.*, 'Stress and Magnetic Field Bimode Detection Sensors Based on Flexible CI/CNTs–PDMS Sponges', *ACS Appl. Mater. Interfaces*, vol. 10, no. 36, pp. 30774–84, 2018.

[190] S. Han, S. Hong, J. Ham *et al.*, 'Fast Plasmonic Laser Nanowelding for a Cu-Nanowire Percolation Network for Flexible Transparent Conductors and Stretchable Electronics', *Adv. Mater.*, vol. 26, no. 33, pp. 5808–14, 2014.

[191] J. Song, J. Li, J. Xu and H. Zeng, 'Superstable Transparent Conductive Cu@Cu4Ni Nanowire Elastomer Composites against Oxidation, Bending, Stretching, and Twisting for Flexible and Stretchable Optoelectronics', *Nano Lett.*, vol. 14, no. 11, pp. 6298–305, 2014.

[192] W. Hu, R. Wang, Y. Lu and Q. Pei, 'An Elastomeric Transparent Composite Electrode Based on Copper Nanowires and Polyurethane', *J. Mater. Chem. C*, vol. 2, no. 7, pp. 1298–305, 2014.

[193] E. O. Polat, O. Balci, N. Kakenov, H. B. Uzlu, C. Kocabas and R. Dahiya, 'Synthesis of Large Area Graphene for High Performance in Flexible Optoelectronic Devices', *Sci. Rep.*, vol. 5, no. 1, p. 16744, 2015.

[194] D. J. Lipomi, M. Vosgueritchian, B. C. K. Tee *et al.*, 'Skin-Like Pressure and Strain Sensors Based on Transparent Elastic Films of Carbon Nanotubes', *Nat. Nanotechnol.*, vol. 6, no. 12, pp. 788–92, 2011.

[195] W. Ma, L. Song, R. Yang *et al.*, 'Directly Synthesized Strong, Highly Conducting, Transparent Single-Walled Carbon Nanotube Films', *Nano Lett.*, vol. 7, no. 8, pp. 2307–11, 2007.

[196] L. Cai, J. Li, P. Luan *et al.*, 'Highly Transparent and Conductive Stretchable Conductors Based on Hierarchical Reticulate Single-Walled Carbon Nanotube Architecture', *Adv. Funct. Mater.*, vol. 22, no. 24, pp. 5238–44, 2012.

[197] F. Xu, X. Wang, Y. Zhu and Y. Zhu, 'Wavy Ribbons of Carbon Nanotubes for Stretchable Conductors', *Adv. Funct. Mater.*, vol. 22, no. 6, pp. 1279–83, 2012.

[198] T. Ann Kim, S.-S. Lee, H. Kim and M. Park, 'Acid-Treated SWCNT/ Polyurethane Nanoweb as a Stretchable and Transparent Conductor', *RSC Adv.*, vol. 2, no. 28, pp. 10717–24, 2012.

[199] B. J. Kim, S.-K. Lee, M. S. Kang, J.-H. Ahn and J. H. Cho, 'Coplanar-Gate Transparent Graphene Transistors and Inverters on Plastic', *ACS Nano*, vol. 6, no. 10, pp. 8646–51, 2012.

[200] N. Yogeswaran, E. S. Hosseini and R. Dahiya, 'Graphene Based Low Voltage Field Effect Transistor Coupled with Biodegradable Piezoelectric Material Based Dynamic Pressure Sensor', *ACS Appl. Mater. Interfaces*, vol. 12, no., no. 48, pp. 54035–40, 2020.

[201] F. Liu, W. T. Navaraj, N. Yogeswaran, D. H. Gregory and R. Dahiya, 'Van der Waals Contact Engineering of Graphene Field-Effect Transistors for Large-Area Flexible Electronics', *ACS Nano*, vol. 13, no. 3, pp. 3257–68, 2019.

[202] S. Bae, H. Kim, Y. Lee *et al.*, 'Roll-to-Roll Production of 30-Inch Graphene Films for Transparent Electrodes', *Nat. Nanotechnol.*, vol. 5, no. 8, pp. 574–78, 2010.

[203] C. García Núñez, W. T. Navaraj, E. O. Polat and R. Dahiya, 'Energy-Autonomous, Flexible, and Transparent Tactile Skin', *Adv. Funct. Mater.*, vol. 27, no. 18, p. 1606287, 2017.

[204] J. Ryu, Y. Kim, D. Won *et al.*, 'Fast Synthesis of High-Performance Graphene Films by Hydrogen-Free Rapid Thermal Chemical Vapor Deposition', *ACS Nano*, vol. 8, no. 1, pp. 950–6, 2014.

[205] R. Dahiya and C. García Núñez, 'Sensor and Devices Incorporating Sensors', patent, PCT/EP2018/054006, 2018.

[206] K. S. Kim, Y. Zhao, H. Jang *et al.*, 'Large-Scale Pattern Growth of Graphene Films for Stretchable Transparent Electrodes', *Nature*, vol. 457, no. 7230, pp. 706–10, 2009.

[207] S.-K. Lee, B. J. Kim, H. Jang *et al.*, 'Stretchable Graphene Transistors with Printed Dielectrics and Gate Electrodes', *Nano Lett.*, vol. 11, no. 11, pp. 4642–6, 2011.

[208] R.-H. Kim, M.-H. Bae, D. G. Kim *et al.*, 'Stretchable, Transparent Graphene Interconnects for Arrays of Microscale Inorganic Light Emitting Diodes on Rubber Substrates', *Nano Lett.*, vol. 11, no. 9, pp. 3881–6, 2011.

[209] Y. Wang, L. Wang, T. Yang *et al.*, 'Wearable and Highly Sensitive Graphene Strain Sensors for Human Motion Monitoring', *Adv. Funct. Mater.*, vol. 24, no. 29, pp. 4666–70, 2014.

[210] C. Wang, W. Zheng, Z. Yue, C. O. Too and G. G. Wallace, 'Buckled, Stretchable Polypyrrole Electrodes for Battery Applications', *Adv. Mater.*, vol. 23, no. 31, pp. 3580–4, 2011.

[211] D. J. Lipomi, J. A. Lee, M. Vosgueritchian, B. C. K. Tee, J. A. Bolander and Z. Bao, 'Electronic Properties of Transparent Conductive Films of PEDOT:PSS on Stretchable Substrates', *Chem. Mater.*, vol. 24, no. 2, pp. 373–82, 2012.

[212] M. Vosgueritchian, D. J. Lipomi and Z. Bao, 'Highly Conductive and Transparent PEDOT:PSS Films with a Fluorosurfactant for Stretchable and Flexible Transparent Electrodes', *Adv. Funct. Mater.*, vol. 22, no. 2, pp. 421–8, 2012.

[213] Y. Wang, C. Zhu, R. Pfattner *et al.*, 'A Highly Stretchable, Transparent, and Conductive Polymer', *Sci. Adv.*, vol. 3, no. 3, p. e1602076, 2017.

[214] J. Y. Oh, M. Shin, J. B. Lee, J.-H. Ahn, H. K. Baik and U. Jeong, 'Effect of PEDOT Nanofibril Networks on the Conductivity, Flexibility, and Coatability of PEDOT:PSS Films', *ACS Appl. Mater. Interfaces*, vol. 6, no. 9, pp. 6954–61, 2014.

[215] M. Soni, M. Bhattacharjee, M. Ntagios and R. Dahiya, 'Printed Temperature Sensor Based on PEDOT:PSS – Graphene Oxide Composite', *IEEE Sens. J.*, vol. 20, no. 14, pp. 7525–31, 2020.

[216] D. Alemu, H.-Y. Wei, K.-C. Ho and C.-W. Chu, 'Highly Conductive PEDOT:PSS Electrode by Simple Film Treatment with Methanol for ITO-Free Polymer Solar Cells', *Energy Environ. Sci.*, vol. 5, no. 11, pp. 9662–71, 2012.

[217] Y. H. Kim, C. Sachse, M. L. Machala, C. May, L. Müller-Meskamp and K. Leo, 'Highly Conductive PEDOT:PSS Electrode with Optimized Solvent and Thermal Post-treatment for ITO-Free Organic Solar Cells', *Adv. Funct. Mater.*, vol. 21, no. 6, pp. 1076–81, 2011.

[218] S. H. Lee, J. S. Sohn, S. B. Kulkarni, U. M. Patil, S. C. Jun and J. H. Kim, 'Modified Physico-chemical Properties and Supercapacitive Performance via DMSO Inducement to PEDOT:PSS Active Layer', *Org. Electron.*, vol. 15, no. 12, pp. 3423–30, 2014.

[219] J. Y. Kim, J. H. Jung, D. E. Lee and J. Joo, 'Enhancement of Electrical Conductivity of poly(3,4-ethylenedioxythiophene)/poly(4-styrenesulfonate) by a Change of Solvents', *Synth. Met.*, vol. 126, no. 2, pp. 311–16, 2002.

[220] J. Ouyang, Q. Xu, C.-W. Chu, Y. Yang, G. Li and J. Shinar, 'On the Mechanism of Conductivity Enhancement in Poly(3,4-ethylenedioxythiophene): Poly(styrenesulfonate) Film through Solvent Treatment', *Polymer*, vol. 45, no. 25, pp. 8443–50, 2004.

[221] R. Dahiya and M. Valle, *Robotic Tactile Sensing: Technologies and System.* Dordrecht: Springer Science Business Media, 2013.

[222] R. Taherian, 'Development of an Equation to Model Electrical Conductivity of Polymer-Based Carbon Nanocomposites', *ECS J. Solid State Sc.*, vol. 3, no. 6, pp. M26–38, 2014.

[223] R. M. Mutiso, M. C. Sherrott, A. R. Rathmell, B. J. Wiley and K. I. Winey, 'Integrating Simulations and Experiments To Predict Sheet Resistance and Optical Transmittance in Nanowire Films for Transparent Conductors', *ACS Nano*, vol. 7, no. 9, pp. 7654–63, 2013.

[224] J. Yang, W. Cheng and K. Kalantar-zadeh, 'Electronic Skins Based on Liquid Metals', *Proc. IEEE*, vol. 107, no. 10, pp. 2168–84, 2019.

[225] N. B. Morley, J. Burris, L. C. Cadwallader and M. D. Nornberg, 'GaInSn Usage in the Research Laboratory', *Rev. Sci. Instrum.*, vol. 79, no. 5, p. 056107, 2008.

[226] C. Pan, K. Kumar, J. Li, E. J. Markvicka, P. R. Herman and C. Majidi, 'Visually Imperceptible Liquid-Metal Circuits for Transparent, Stretchable Electronics with Direct Laser Writing', *Adv. Mater.*, vol. 30, no. 12, p. 1706937, 2018.

[227] Y. Yang, N. Sun, Z. Wen *et al.*, 'Liquid-Metal-Based Super-Stretchable and Structure-Designable Triboelectric Nanogenerator for Wearable Electronics', *ACS Nano*, vol. 12, no. 2, pp. 2027–34, 2018.

[228] M. D. Dickey, R. C. Chiechi, R. J. Larsen, E. A. Weiss, D. A. Weitz and G. M. Whitesides, 'Eutectic Gallium-Indium (EGaIn): A Liquid Metal Alloy for the Formation of Stable Structures in Microchannels at Room Temperature', *Adv. Funct. Mater.*, vol. 18, no. 7, pp. 1097–104, 2008.

[229] C. Ladd, J.-H. So, J. Muth and M. D. Dickey, '3D Printing of Free Standing Liquid Metal Microstructures', *Adv. Mater.*, vol. 25, no. 36, pp. 5081–5, 2013.

[230] S. H. Jeong, A. Hagman, K. Hjort, M. Jobs, J. Sundqvist and Z. Wu, 'Liquid Alloy Printing of Microfluidic Stretchable Electronics', *Lab Chip*, vol. 12, no. 22, pp. 4657–64, 2012.

[231] S. Cheng and Z. Wu, 'Microfluidic Stretchable RF Electronics', *Lab Chip*, vol. 10, no. 23, pp. 3227–34, 2010.

[232] S. H. Jeong, K. Hjort and Z. Wu, 'Tape Transfer Printing of a Liquid Metal Alloy for Stretchable RF Electronics', *Sensors*, vol. 14, no. 9, pp. 16311–21, 2014.

[233] T. Lu, L. Finkenauer, J. Wissman and C. Majidi, 'Rapid Prototyping for Soft-Matter Electronics', *Adv. Funct. Mater.*, vol. 24, no. 22, pp. 3351–6, 2014.

[234] S. Zhu, J.-H. So, R. Mays *et al.*, 'Ultrastretchable Fibers with Metallic Conductivity Using a Liquid Metal Alloy Core', *Adv. Funct. Mater.*, vol. 23, no. 18, pp. 2308–14, 2013.

[235] E. Palleau, S. Reece, S. C. Desai, M. E. Smith and M. D. Dickey, 'Self-healing Stretchable Wires for Reconfigurable Circuit Wiring and 3D Microfluidics', *Adv. Mater.*, vol. 25, no. 11, pp. 1589–92, 2013.

[236] A. Vilouras, A. Christou, L. Manjakkal and R. Dahiya, 'Ultrathin Ion-Sensitive Field-Effect Transistor Chips with Bending-Induced Performance Enhancement', *ACS Applied Electronic Materials*, vol. 2, no. 8, pp. 2601–10, 2020.

[237] Y. Kumaresan, H. Kim, Y. Pak *et al.*, 'Omnidirectional Stretchable Inorganic-Material-Based Electronics with Enhanced Performance', *Adv. Electron. Mater.*, vol. 6, no. 7, p. 2000058, 2020.

[238] S. Savagatrup, A. D. Printz, H. Wu *et al.*, 'Viability of Stretchable Poly (3-heptylthiophene) (P3HpT) for Organic Solar Cells and Field-Effect Transistors', *Synth. Met.*, vol. 203, pp. 208–14, 2015.

[239] T. Sekitani, Y. Noguchi, K. Hata, T. Fukushima, T. Aida and T. Someya, 'A Rubberlike Stretchable Active Matrix Using Elastic Conductors', *Science*, vol. 321, no. 5895, p. 1468, 2008.

[240] E. Song, B. Kang, H. H. Choi *et al.*, 'Stretchable and Transparent Organic Semiconducting Thin Film with Conjugated Polymer Nanowires Embedded in an Elastomeric Matrix', *Adv. Electron. Mater.*, vol. 2, no. 1, p. 1500250, 2016.

[241] C. Müller, S. Goffri, D. W. Breiby *et al.*, 'Tough, Semiconducting Polyethylene-poly(3-hexylthiophene) Diblock Copolymers', *Adv. Funct. Mater.*, vol. 17, no. 15, pp. 2674–9, 2007.

[242] G.-J. N. Wang, L. Shaw, J. Xu *et al.*, 'Inducing Elasticity through Oligo-Siloxane Crosslinks for Intrinsically Stretchable Semiconducting Polymers', *Adv. Funct. Mater.*, vol. 26, no. 40, pp. 7254–62, 2016.

[243] J. Y. Oh, S. Rondeau-Gagné, Y.-C. Chiu *et al.*, 'Intrinsically Stretchable and Healable Semiconducting Polymer for Organic Transistors', *Nature*, vol. 539, no. 7629, pp. 411–15, 2016.

[244] J. M. London, A. H. Loomis, J. F. Ahadian and C. G. Fonstad, 'Preparation of Silicon-on-Gallium Arsenide Wafers for Monolithic Optoelectronic Integration', *IEEE Photonics Tech. L.*, vol. 11, no. 8, pp. 958–60, 1999.

[245] H. Xu, L. Yin, C. Liu, X. Sheng and N. Zhao, 'Recent Advances in Biointegrated Optoelectronic Devices', *Adv. Mater.*, vol. 30, no. 33, p. 1800156, 2018.

[246] F. Liu, A. S. Dahiya and R. Dahiya, 'A Flexible Chip with Embedded Intelligence', *Nat. Electron.*, vol. 3, no. 7, pp. 358–9, 2020.

[247] W. T. Navaraj, S. Gupta, L. Lorenzelli and R. Dahiya, 'Wafer Scale Transfer of Ultrathin Silicon Chips on Flexible Substrates for High Performance Bendable Systems', *Adv. Electron. Mater.*, vol. 4, no. 4, p. 1700277, 2018.

[248] S. Gupta, W. T. Navaraj, L. Lorenzelli and R. Dahiya, 'Ultra-Thin Chips for High-Performance Flexible Electronics', *npj Flex. Electron.*, vol. 2, no. 1, p. 8, 2018.

[249] R. S. Dahiya and S. Gennaro, 'Bendable Ultra-Thin Chips on Flexible Foils', *IEEE Sens. J.*, vol. 13, no. 10, pp. 4030–7, 2013.

[250] Y. K. Lee, K. J. Yu, E. Song *et al.*, 'Dissolution of Monocrystalline Silicon Nanomembranes and Their Use as Encapsulation Layers and Electrical Interfaces in Water-Soluble Electronics', *ACS Nano*, vol. 11, no. 12, pp. 12562–72, 2017.

[251] R. Dahiya, G. Gottardi and N. Laidani, 'PDMS Residues-Free Micro/macrostructures on Flexible Substrates', *Microelectron. Eng.*, vol. 136, pp. 57–62, 2015.

[252] R. S. Dahiya, A. Adami, C. Collini and L. Lorenzelli, 'Fabrication of Single Crystal Silicon Micro-/nanostructures and Transferring Them to Flexible Substrates', *Microelectron. Eng.*, vol. 98, pp. 502–7, 2012.

[253] D. Shakthivel, W. T. Navaraj, S. Champet, D. H. Gregory and R. S. Dahiya, 'Propagation of Amorphous Oxide Nanowires via the VLS Mechanism: Growth Kinetics', *Nanoscale Adv.*, vol. 1, no. 9, pp. 3568–78, 2019.

[254] A. Ejaz, J. H. Han and R. Dahiya, 'Influence of Solvent Molecular Geometry on the Growth of Nanostructures', *J. Colloid Interface Sci.*, vol. 570, pp. 322–31, 2020.

[255] D. Shakthivel, M. Ahmad, M. R. Alenezi, R. Dahiya and S. R. P. Silva, *1D Semiconducting Nanostructures for Flexible and Large-Area Electronics: Growth Mechanisms and Suitability.* Cambridge: Cambridge University Press, 2019. www.cambridge.org/core/elements/1d-semiconducting-nanostructures-for-flexible-and-largearea-electronics/AB256ABD4C286D6E270A8021CFD930FE.

[256] C. García Núñez, F. Liu, S. Xu and R. Dahiya, *Integration Techniques for Micro/Nanostructure- Based Large-Area Electronics.* Cambridge: Cambridge University Press, 2018. www.cambridge.org/core/elements/integration-techniques-for- micronanostructurebased-largearea-electronics/F83E7CAF7CDBE93E69C7434C3EE29DED.

[257] A. Zumeit, W. T. Navaraj, D. Shakthivel and R. Dahiya, 'Nanoribbon-Based Flexible High-Performance Transistors Fabricated at Room Temperature', *Adv. Electron. Mater.*, vol. 6, no. 4, p. 1901023, 2020.

[258] D.-Y. Khang, H. Jiang, Y. Huang and J. A. Rogers, 'A Stretchable Form of Single-Crystal Silicon for High-Performance Electronics on Rubber Substrates', *Science*, vol. 311, no. 5758, p. 208, 2006.

[259] C. García Núñez, W. T. Navaraj, F. Liu, D. Shakthivel and R. Dahiya, 'Large-Area Self-Assembly of Silica Microspheres/Nanospheres by Temperature-Assisted Dip-Coating', *ACS Appl. Mater. Interfaces*, vol. 10, no. 3, pp. 3058–68, 2018.

[260] Y. Sun, W. M. Choi, H. Jiang, Y. Y. Huang and J. A. Rogers, 'Controlled Buckling of Semiconductor Nanoribbons for Stretchable Electronics', *Nat. Nanotechnol.*, vol. 1, no. 3, pp. 201–7, 2006.

[261] A. S. Dahiya, Y. Kumaresan, D. Shakthivel, A. Zumeit, A. Christou and R. Dahiya, 'High-Performance Printed Electronics based on Inorganic Semiconducting Nano to Chip Scale Structures' *Nano Converg.*, vol. 7, no.NoNo. 37, 2020.

[262] F. Xu, M.-Y. Wu, N. S. Safron *et al.*, 'Highly Stretchable Carbon Nanotube Transistors with Ion Gel Gate Dielectrics', *Nano Lett.*, vol. 14, no. 2, pp. 682–6, 2014.

[263] S. H. Chae, W. J. Yu, J. J. Bae *et al.*, 'Transferred Wrinkled Al₂O₃ for Highly Stretchable and Transparent Graphene–Carbon Nanotube Transistors', *Nat. Mater.*, vol. 12, no. 5, pp. 403–9, 2013.

[264] X. Yu, K. B. Mahajan, W. Shou and H. Pan, 'Materials, Mechanics, and Patterning Techniques for Elastomer-Based Stretchable Conductors', *Micromachines*, vol. 8, no. 1, p. 7, 2017.

[265] H. Miyajima and M. Mehregany, 'High-Aspect-Ratio Photolithography for MEMS Applications', *J. Microelectromech. S.*, vol. 4, no. 4, pp. 220–9, 1995.

[266] C. L. Tuinea-Bobe, P. Lemoine, M. U. Manzoor *et al.*, 'Photolithographic Structuring of Stretchable Conductors and Sub-kPa Pressure Sensors', *J. Micromech.Microeng.*, vol. 21, no. 11, p. 115010, 2011.

[267] C. Acikgoz, M. A. Hempenius, J. Huskens and G. J. Vancso, 'Polymers in Conventional and Alternative Lithography for the Fabrication of Nanostructures', *Eur. Polym. J.*, vol. 47, no. 11, pp. 2033–52, 2011.

[268] L. Guo and S. P. DeWeerth, 'An Effective Lift-Off Method for Patterning High-Density Gold Interconnects on an Elastomeric Substrate', *Small*, vol. 6, no. 24, pp. 2847–52, 2010.

[269] J. N. Patel, B. Kaminska, B. L. Gray and B. D. Gates, 'A Sacrificial SU-8 Mask for Direct Metallization on PDMS', *J. Micromech.Microeng.*, vol. 19, no. 11, p. 115014, 2009.

[270] M.-G. Kang and L. J. Guo, 'Metal Transfer Assisted Nanolithography on Rigid and Flexible Substrates', *J. Vac. Sci. Technol. B*, vol. 26, no. 6, pp. 2421–5, 2008.

[271] T. Adrega and S. P. Lacour, 'Stretchable Gold Conductors Embedded in PDMS and Patterned by Photolithography: Fabrication and Electromechanical Characterization', *J. Micromech.Microeng.*, vol. 20, no. 5, p. 055025, 2010.

[272] C. Tsay, S. P. Lacour, S. Wagner and B. Morrison, 'Architecture, Fabrication, and Properties of Stretchable Micro-electrode Arrays', *SENSORS, 2005 IEEE*, 2005, pp. 1169–72.

[273] S. P. Lacour, D. Chan, S. Wagner, T. Li and Z. Suo, 'Mechanisms of Reversible Stretchability of Thin Metal Films on Elastomeric Substrates', *Appl. Phys. Lett.*, vol. 88, no. 20, p. 204103, 2006.

[274] A. J. Bandodkar, R. Nuñez-Flores, W. Jia and J. Wang, 'All-printed Stretchable Electrochemical Devices', *Adv. Mater.*, vol. 27, no. 19, pp. 3060–5, 2015.

[275] S. Yao and Y. Zhu, 'Wearable Multifunctional Sensors Using Printed Stretchable Conductors Made of Silver Nanowires', *Nanoscale*, vol. 6, no. 4, pp. 2345–52, 2014.

[276] S. Zhang, E. Hubis, G. Tomasello, G. Soliveri, P. Kumar and F. Cicoira, 'Patterning of Stretchable Organic Electrochemical Transistors', *Chem. Mater.*, vol. 29, no. 7, pp. 3126–32, 2017.

[277] S. Chung, J. Lee, H. Song, S. Kim, J. Jeong and Y. Hong, 'Inkjet-Printed Stretchable Silver Electrode on Wave Structured Elastomeric Substrate', *Appl. Phys. Lett.*, vol. 98, no. 15, p. 153110, 2011.

[278] Y. Kim, X. Ren, J. W. Kim and H. Noh, 'Direct Inkjet Printing of Micro-scale Silver Electrodes on Polydimethylsiloxane (PDMS) Microchip', *J. Micromech.Microeng.*, vol. 24, no. 11, p. 115010, 2014.

[279] X.-M. Zhao, Y. Xia and G. M. Whitesides, 'Soft Lithographic Methods for Nano-fabrication', *J. Mater. Chem.*, vol. 7, no. 7, pp. 1069–74, 1997.

[280] M. Husemann, D. Mecerreyes, C. J. Hawker, J. L. Hedrick, R. Shah and N. L. Abbott, 'Surface- Initiated Polymerization for Amplification of Self-assembled Monolayers Patterned by Microcontact Printing', *Angew. Chem.*, vol. 38, no. 5, pp. 647–9, 1999.

[281] M. Toprak, D. K. Kim, M. Mikhailova and M. Muhammed, 'Patterning 2D Metallic Surfaces by Soft Lithography', *MRS Proceedings*, vol. 705, Y7.22, 2001.

[282] Y.-L. Loo, R. L. Willett, K. W. Baldwin and J. A. Rogers, 'Additive, Nanoscale Patterning of Metal Films with a Stamp and a Surface Chemistry Mediated Transfer Process: Applications in Plastic Electronics', *Appl. Phys. Lett.*, vol. 81, no. 3, pp. 562–4, 2002.

[283] E. K. W. Tan, 'Technological Development of Chemical Sensors for Healthcare and Safety Applications', doctorate, University of Cambridge, 2019.

[284] E. K. W. Tan, G. Rughoobur, J. Rubio-Lara *et al.*, 'Nanofabrication of Conductive Metallic Structures on Elastomeric Materials', *Sci. Rep.*, vol. 8, no. 1, p. 6607, 2018.

[285] X. Wen, G. Li, J. Zhang *et al.*, 'Transparent Free-standing Metamaterials and Their Applications in Surface- Enhanced Raman Scattering', *Nanoscale*, vol. 6, no. 1, pp. 132–9, 2014.

[286] A. D. Valentine, T. A. Busbee, J. W. Boley *et al.*, 'Hybrid 3D Printing of Soft Electronics', *Adv. Mater.*, vol. 29, no. 40, p. 1703817, 2017.

[287] O. Ozioko, H. Nassar and R. Dahiya, '3D Printed Interdigitated Capacitors based Tilt Sensor', *IEEE Sens. J.*, 2021 (DOI:10.1109/JSEN.2021.3058949).

[288] T. Distler and A. R. Boccaccini, '3D Printing of Electrically Conductive Hydrogels for Tissue Engineering and Biosensors – A Review', *Acta Biomater.*, vol. 101, pp. 1–13, 2020.

[289] M. Abshirini, M. Charara, Y. Liu, M. Saha and M. C. Altan, '3D Printing of Highly Stretchable Strain Sensors Based on Carbon Nanotube Nanocomposites', *Adv. Eng. Mater.*, vol. 20, no. 10, p. 1800425, 2018.

[290] M. Mohammed Ali, D. Maddipatla, B. B. Narakathu *et al.*, 'Printed Strain Sensor Based on Silver Nanowire/Silver Flake Composite on Flexible and Stretchable TPU Substrate', *Sens. Actuator A Phys.*, vol. 274, pp. 109–15, 2018.

[291] J. F. Salmerón, F. Molina-Lopez, D. Briand *et al.*, 'Properties and Printability of Inkjet and Screen-Printed Silver Patterns for RFID Antennas', *J. Electron. Mater.*, vol. 43, no. 2, pp. 604–17, 2014.

[292] X. Cao, C. Lau, Y. Liu *et al.*, 'Fully Screen-Printed, Large-Area, and Flexible Active-Matrix Electrochromic Displays Using Carbon Nanotube Thin-Film Transistors', *ACS Nano*, vol. 10, no. 11, pp. 9816–22, 2016.

[293] A. J. Bandodkar, I. Jeerapan, J.-M. You, R. Nuñez-Flores and J. Wang, 'Highly Stretchable Fully- Printed CNT-Based Electrochemical Sensors and Biofuel Cells: Combining Intrinsic and Design-Induced Stretchability', *Nano Lett.*, vol. 16, no. 1, pp. 721–7, 2016.

[294] Y. Tong, Z. Feng, J. Kim, J. L. Robertson, X. Jia and B. N. Johnson, '3D Printed Stretchable Triboelectric Nanogenerator Fibers and Devices', *Nano Energy*, vol. 75, p. 104973, 2020.

[295] S. Peng, Y. Li, L. Wu *et al.*, '3D Printing Mechanically Robust and Transparent Polyurethane Elastomers for Stretchable Electronic Sensors', *ACS Appl. Mater. Interfaces*, vol. 12, no. 5, pp. 6479–88, 2020.

[296] J. Pu, X. Wang, R. Xu and K. Komvopoulos, 'Highly Stretchable Microsupercapacitor Arrays with Honeycomb Structures for Integrated Wearable Electronic Systems', *ACS Nano*, vol. 10, no. 10, pp. 9306–15, 2016.

[297] D. J. Lipomi, 'Stretchable Figures of Merit in Deformable Electronics', *Adv. Mater.*, vol. 28, no. 22, pp. 4180–3, 2016.

[298] Y. Liu, M. Pharr and G. A. Salvatore, 'Lab-on-Skin: A Review of Flexible and Stretchable Electronics for Wearable Health Monitoring', *ACS Nano*, vol. 11, no. 10, pp. 9614–35, 2017.

[299] X. Yang and H. Cheng, 'Recent Developments of Flexible and Stretchable Electrochemical Biosensors', *Micromachines*, vol. 11, no. 3, 2020.

[300] W. Gao, S. Emaminejad, H. Y. Y. Nyein *et al.*, 'Fully Integrated Wearable Sensor Arrays for Multiplexed In Situ Perspiration Analysis', *Nature*, vol. 529, no. 7587, pp. 509–14, 2016.

[301] A. Koh, D. Kang, Y. Xue *et al.*, 'A Soft, Wearable Microfluidic Device for the Capture, Storage, and Colorimetric Sensing of Sweat', *Sci. Transl. Med.*, vol. 8, no. 366, p. 366ra165, 2016.

[302] G. Wang, S. Zhang, S. Dong *et al.*, 'Stretchable Optical Sensing Patch System Integrated Heart Rate, Pulse Oxygen Saturation, and Sweat pH Detection', *IEEE. Trans. Biomed. Eng.*, vol. 66, no. 4, pp. 1000–5, 2019.

[303] A. A. Al-Halhouli, L. Al-Ghussain, S. El Bouri, H. Liu and D. Zheng, 'Fabrication and Evaluation of a Novel Non-invasive Stretchable and Wearable Respiratory Rate Sensor Based on Silver Nanoparticles Using Inkjet Printing Technology', *Polymers*, vol. 11, no. 9, p. 1518, 2019.

[304] S. Y. Oh, S. Y. Hong, Y. R. Jeong *et al.*, 'Skin-Attachable, Stretchable Electrochemical Sweat Sensor for Glucose and pH Detection', *ACS Appl. Mater. Interfaces*, vol. 10, no. 16, pp. 13729–40, 2018.

[305] N. Matsuhisa, M. Kaltenbrunner, T. Yokota *et al.*, 'Printable Elastic Conductors with a High Conductivity for Electronic Textile Applications', *Nat. Commun.*, vol. 6, no. 1, p. 7461, 2015.

[306] Y. Jiang, Z. Liu, N. Matsuhisa *et al.*, 'Auxetic Mechanical Metamaterials to Enhance Sensitivity of Stretchable Strain Sensors', *Adv. Mater.*, vol. 30, no. 12, p. 1706589, 2018.

[307] J. Zou, M. Zhang, J. Huang *et al.*, 'Coupled Supercapacitor and Triboelectric Nanogenerator Boost Biomimetic Pressure Sensor', *Adv. Energy Mater.*, vol. 8, no. 10, p. 1702671, 2018.

[308] S. Gupta, D. Shakthivel, L. Lorenzelli and R. Dahiya, 'Temperature Compensated Tactile Sensing Using MOSFET With P(VDF-TrFE)/BaTiO$_3$ Capacitor as Extended Gate', *IEEE Sens. J.*, vol. 19, no. 2, pp. 435–42, 2019.

[309] H. Kim, W. Kim, J. Park *et al.*, 'Surface Conversion of ZnO Nanorods to ZIF-8 to Suppress Surface Defects for a Visible-Blind UV Photodetector', *Nanoscale*, vol. 10, no. 45, pp. 21168–77, 2018.

[310] L.-B. Luo, D. Wang, C. Xie, J.-G. Hu, X.-Y. Zhao and F.-X. Liang, 'PdSe2 Multilayer on Germanium Nanocones Array with Light Trapping Effect for Sensitive Infrared Photodetector and Image Sensing Application', *Adv. Funct. Mater.*, vol. 29, no. 22, p. 1900849, 2019.

[311] Y. Dai, H. Hu, M. Wang, J. Xu and S. Wang, 'Stretchable Transistors and Functional Circuits for Human-Integrated Electronics', *Nat. Electron.*, vol. 4, no. 1, pp. 17–29, 2021.

[312] Y. Guo, Y. Li, Q. Zhang and H. Wang, 'Self-powered Multifunctional UV and IR Photodetector as an Artificial Electronic Eye', *J. Mater. Chem. C*, vol. 5, no. 6, pp. 1436–42, 2017.

[313] W. Deng, X. Zhang, L. Huang *et al.*, 'Aligned Single-Crystalline Perovskite Microwire Arrays for High-Performance Flexible Image Sensors with Long-Term Stability', *Adv. Mater.*, vol. 28, no. 11, pp. 2201–8, 2016.

[314] W. Lee, J. Lee, H. Yun *et al.*, 'High-Resolution Spin-on-Patterning of Perovskite Thin Films for a Multiplexed Image Sensor Array', *Adv. Mater.*, vol. 29, no. 40, p. 1702902, 2017.

[315] J. Kim, H. Park and S.-H. Jeong, 'A Kirigami Concept for Transparent and Stretchable Nanofiber Networks-Based Conductors and UV Photodetectors', *J. Ind. Eng. Chem.*, vol. 82, pp. 144–52, 2020.

[316] P. Kang, M. C. Wang, P. M. Knapp and S. Nam, 'Crumpled Graphene Photodetector with Enhanced, Strain-Tunable, and Wavelength-Selective Photoresponsivity', *Adv. Mater.*, vol. 28, no. 23, pp. 4639–45, 2016.

[317] M. Kim, P. Kang, J. Leem and S. Nam, 'A Stretchable Crumpled Graphene Photodetector with Plasmonically Enhanced Photoresponsivity', *Nanoscale*, vol. 9, no. 12, pp. 4058–65, 2017.

[318] P. Liu, X. He, J. Ren, Q. Liao, J. Yao and H. Fu, 'Organic–Inorganic Hybrid Perovskite Nanowire Laser Array', *ACS Nano*, vol. 11, p. 8, 2017.

[319] B. P. Yalagala, P. Sahatiya, C. S. R. Kolli, S. Khandelwal, V. Mattela and S. Badhulika, 'V_2O_5 Nanosheets for Flexible Memristors and Broadband Photodetectors', *ACS Appl. Nano Mater.*, vol. 2, no. 2, pp. 937–47, 2019.

[320] L. Li, L. Gu, Z. Lou, Z. Fan and G. Shen, 'ZnO Quantum Dot Decorated Zn2SnO4 Nanowire Heterojunction Photodetectors with Drastic Performance Enhancement and Flexible Ultraviolet Image Sensors', *ACS Nano*, vol. 11, no. 4, pp. 4067–76, 2017.

[321] J. Yu, K. Javaid, L. Liang *et al.*, 'High-Performance Visible-Blind Ultraviolet Photodetector Based on IGZO TFT Coupled with p–n Heterojunction', *ACS Appl. Mater. Interfaces*, vol. 10, no. 9, pp. 8102–9, 2018.

[322] A. Yao, Z. Luo, P. Yin *et al.*, 'Formation and Applications of Highly-Ordered CdO Nanobranch Arrays', *Mater. Lett.*, vol. 172, pp. 132–6, 2016.

[323] L. Manjakkal, D. Szwagierczak and R. Dahiya, 'Metal Oxides Based Electrochemical pH Sensors: Current Progress and Future Perspectives', *Prog. Mater. Sci.*, vol. 109, p. 100635, 2020.

[324] N. Mathews, B. Varghese, C. Sun *et al.*, 'Oxide Nanowire Networks and Their Electronic and Optoelectronic Characteristics', *Nanoscale*, vol. 2, no. 10, pp. 1984–98, 2010.

[325] M. Bhattacharjee, F. Nikbakhtnasrabadi and R. Dahiya, 'Printed Chipless Antenna as Flexible Temperature Sensor', *IEEE Internet Things J.*, vol. 8, no. 6, pp. 5101–10, 2021.

[326] O. Ozioko, P. Karipoth, P. Escobedo, M. Ntagios, A. Pullanchiyodan and R. Dahiya, 'SensAct: The Soft and Squishy Tactile Sensor with Integrated Flexible Actuator', *Adv. Intell. Syst.*, vol. 3, no. 3, 1900145, 2021.

[327] G. Wang, Z. Wang, Y. Wu *et al.*, 'A Robust Stretchable Pressure Sensor for Electronic Skins', *Org. Electron.*, vol. 86, p. 105926, 2020.

[328] M. Ntagios, H. Nassar, A. Pullanchiyodan, W. T. Navaraj and R. Dahiya, 'Robotic Hands with Intrinsic Tactile Sensing via 3D Printed Soft Pressure Sensors', *Adv. Intell. Syst.*, vol. 2, 1900080, 2020.

[329] H.-J. Kim, A. Thukral and C. Yu, 'Highly Sensitive and Very Stretchable Strain Sensor Based on a Rubbery Semiconductor', *ACS Appl. Mater. Interfaces*, vol. 10, no. 5, pp. 5000–6, 2018.

[330] G. Choi, S. Oh, C. Kim *et al.*, 'Omnidirectionally Stretchable Organic Transistors for Use in Wearable Electronics: Ensuring Overall

Stretchability by Applying Nonstretchable Wrinkled Components', *ACS Appl. Mater. Interfaces*, vol. 12, no. 29, pp. 32979–86, 2020.

[331] J. Zhao, T. Bu, X. Zhang *et al.*, 'Intrinsically Stretchable Organic-Tribotronic-Transistor for Tactile Sensing', *Research*, vol. 2020, p. 1398903, 2020.

[332] H.-H. Chou, A. Nguyen, A. Chortos *et al.*, 'A Chameleon-Inspired Stretchable Electronic Skin with Interactive Colour Changing Controlled by Tactile Sensing', *Nat. Commun.*, vol. 6, no. 1, p. 8011, 2015.

[333] G. Schwartz, B. C. K. Tee, J. Mei *et al.*, 'Flexible Polymer Transistors with High Pressure Sensitivity for Application in Electronic Skin and Health Monitoring', *Nat. Commun.*, vol. 4, no. 1, p. 1859, 2013.

[334] J. Wang, J. Jiang, C. Zhang *et al.*, 'Energy-Efficient, Fully Flexible, High-Performance Tactile Sensor Based on Piezotronic Effect: Piezoelectric Signal Amplified with Organic Field-Effect Transistors', *Nano Energy*, vol. 76, p. 105050, 2020.

[335] S. Kang, S. Y. Hong, N. Kim *et al.*, 'Stretchable Lithium-Ion Battery Based on Re-entrant Micro-honeycomb Electrodes and Cross-Linked Gel Electrolyte', *ACS Nano*, vol. 14, no. 3, pp. 3660–8, 2020.

[336] Z. Song, T. Ma, R. Tang *et al.*, 'Origami Lithium-Ion Batteries', *Nat. Commun.*, vol. 5, no. 1, p. 3140, 2014.

[337] S. Xu, Y. Zhang, J. Cho *et al.*, 'Stretchable Batteries with Self-similar Serpentine Interconnects and Integrated Wireless Recharging Systems', *Nat. Commun.*, vol. 4, no. 1, p. 1543, 2013.

[338] Z. Song, X. Wang, C. Lv *et al.*, 'Kirigami-based Stretchable Lithium-Ion Batteries', *Sci. Rep.*, vol. 5, no. 1, p. 10988, 2015.

[339] W. Liu, Z. Chen, G. Zhou *et al.*, '3D Porous Sponge-Inspired Electrode for Stretchable Lithium-Ion Batteries', *Adv. Mater.*, vol. 28, no. 18, pp. 3578–83, 2016.

[340] W. Liu, J. Chen, Z. Chen *et al.*, 'Stretchable Lithium-Ion Batteries Enabled by Device-Scaled Wavy Structure and Elastic-Sticky Separator', *Adv. Energy Mater.*, vol. 7, no. 21, p. 1701076, 2017.

[341] K. Liu, B. Kong, W. Liu *et al.*, 'Stretchable Lithium Metal Anode with Improved Mechanical and Electrochemical Cycling Stability', *Joule*, vol. 2, no. 9, pp. 1857–65, 2018.

[342] G. Lee, D. Kim, D. Kim *et al.*, 'Fabrication of a Stretchable and Patchable Array of High Performance Micro- supercapacitors Using a Non-aqueous Solvent Based Gel Electrolyte', *Energy Environ. Sci.*, vol. 8, no. 6, pp. 1764–74, 2015.

[343] H. Guo, M.-H. Yeh, Y.-C. Lai *et al.*, 'All-in-One Shape-Adaptive Self-charging Power Package for Wearable Electronics', *ACS Nano*, vol. 10, no. 11, pp. 10580–8, 2016.

[344] J. Xu, J. Chen, M. Zhang, J.-D. Hong and G. Shi, 'Highly Conductive Stretchable Electrodes Prepared by In Situ Reduction of Wavy Graphene Oxide Films Coated on Elastic Tapes', *Adv. Electron. Mater.*, vol. 2, no. 6, p. 1600022, 2016.

[345] V. Rajendran, A. M. V. Mohan, M. Jayaraman and T. Nakagawa, 'All-printed, Interdigitated, Freestanding Serpentine Interconnects Based Flexible Solid State Supercapacitor for Self Powered Wearable Electronics', *Nano Energy*, vol. 65, p. 104055, 2019.

[346] J. Zang, C. Cao, Y. Feng, J. Liu and X. Zhao, 'Stretchable and High-Performance Supercapacitors with Crumpled Graphene Papers', *Sci. Rep.*, vol. 4, no. 1, p. 6492, 2014.

[347] H. Xiao, Z.-S. Wu, F. Zhou *et al.*, 'Stretchable Tandem Micro-supercapacitors with High Voltage Output and Exceptional Mechanical Robustness', *Energy Storage Mater.*, vol. 13, pp. 233–40, 2018.

[348] J. Yun, Y. Lim, G. N. Jang *et al.*, 'Stretchable Patterned Graphene Gas Sensor Driven by Integrated Micro- supercapacitor Array', *Nano Energy*, vol. 19, pp. 401–14, 2016.

[349] J. Yun, C. Song, H. Lee *et al.*, 'Stretchable Array of High-Performance Micro-supercapacitors Charged with Solar Cells for Wireless Powering of an Integrated Strain Sensor', *Nano Energy*, vol. 49, pp. 644–54, 2018.

Cambridge Elements ☰

Flexible and Large-Area Electronics

Ravinder Dahiya
University of Glasgow

Ravinder Dahiya is a Professor of Electronic and Nanoengineering and EPSRC Fellow at the University of Glasgow. He is the President (2022–23) of the IEEE Sensors Council, the founding Editor-in-Chief of IEEE Journal on Flexible Electronics, and the founder of the IEEE International Conference on Flexible, Printable Sensors and Systems (FLEPS). He is a Distinguished Lecturer of the IEEE Sensors Council, and a Fellow of the IEEE.

Luigi G. Occhipinti
University of Cambridge

Luigi G. Occhipinti is Director of Research at the University of Cambridge, Engineering Department, and Deputy Director and COO of the Cambridge Graphene Centre. He is Founder and CEO at Cambridge Innovation Technologies Consulting Limited, providing research and innovation within both the health care and medical fields. He is a recognised expert in printed, organic, and large-area electronics and integrated smart systems with over 20 years' experience in the semiconductor industry, and is a former R&D Senior Group Manager and Programs Director at STMicroelectronics.

About the Series

This innovative series provides authoritative coverage of the state of the art in bendable and large-area electronics. Specific Elements provide in-depth coverage of key technologies, materials and techniques for the design and manufacturing of flexible electronic circuits and systems, as well as cutting-edge insights into emerging real-world applications. This series is a dynamic reference resource for graduate students, researchers, and practitioners in electrical engineering, physics, chemistry and materials.

Cambridge Elements $^{\equiv}$

Flexible and Large-Area Electronics

Printed in the United States
by Baker & Taylor Publisher Services